3rd International Workshop on Flexibility in Sustainable Construction (ORSDCE 2018)

MATEC Web of Conferences Volume 222 (2018)

Poznan, Poland
20 April 2018

Editor:

J. Paslawski

ISBN: 978-1-5108-7407-7

Printed from e-media with permission by:

Curran Associates, Inc.
57 Morehouse Lane
Red Hook, NY 12571

Some format issues inherent in the e-media version may also appear in this print version.

Printed by Curran Associates, Inc. (2018)

For additional information, please contact EDP Sciences – Web of Conferences at the address below.

EDP Sciences – Web of Conferences
17, Avenue du Hoggar
Parc d'Activité de Courtabœuf
BP 112
F-91944 Les Ulis Cedex A
France

Phone: +33 (0) 1 69 18 75 75
Fax: +33 (0) 1 69 28 84 91

contact-edps@webofconferences.org

Additional copies of this publication are available from:

Curran Associates, Inc.
57 Morehouse Lane
Red Hook, NY 12571 USA
Phone: 845-758-0400
Fax: 845-758-2633
Email: curran@proceedings.com
Web: www.proceedings.com

TABLE OF CONTENTS

ORSDCE Workshop 2018

3rd International Workshop on flexibility in sustainable construction. April, 20th, Poznan

Venue – Poznan University of Technology, PL- 60-965 Poznan, 2 Piotrowo Str., Lecture Center, auditorium no. 25 Technical Library

General aims of the workshop:
– inspiration of young researchers to work grace to keynote speakers
– discussion and development of new ideas and solutions with partners from industry,
– presentation of innovative projects given by the enterprises,
– knowledge bridge creation between academia – industry – PhD students,
– verification of the concepts in practice – case studies in progress

Topics:
flexible management; lean management; BIM; agile management; operational research; decision making; civil engineering; construction technology and management; sustainable development in civil engineering; buildings life cycle; qualification of human resources; real estate management; project management; quality management.

Honorary Committee
O. Kaplinski (Poland)
F. Peldschus (Germany)
E.K. Zavadskas (Lithuania)

Honorary Patronage
T. Lodygowski – Rector of PUT
Section of Construction Project Engineering Committee Civil Engineering of Polish Academy of Science

Organising committee
M. Gajzler (Poland)
P. Nowotarski (Poland)
Construction Innovations Scientific Association

Contact with us:
conference@ztiob.edu.pl
Phone: +48 61 665 2190
www.scfm.put.poznan.pl

International Programme Committee
G. Girmscheid (Switzerland)
J. Jasiczak (Poland)
A. Kaklauskas (Lithuania)
W. Maliene (United Kingdom)
J. Paslawski (Poland)
M. Radujkovič (Croatia)
A. Sobotka (Poland)
J. Šelih (Slovenia)
Z. Turskis (Lithuania)
L. Ustinovičius (Lithuania)
T. Vilutiene (Lithuania)
T. Mróz (Poland)
J.Tamošaitienė (Lithuania)

https://doi.org/10.1051/matecconf/201822200001

Dear Readers,

Dear Partners and Friends,

I am very happy to hand you over proceedings from our 3rd International Workshop on flexibility in sustainable construction. The flexible approach mentioned in the title refers to the current management trends (Industry 4.0, Quality 4.0, Marketing 4.0, and Project Management 2.0) that facilitate the implementation of construction processes in a changing environment based on flexibility options. Naturally, the theme of the speeches covers a broader range of issues related also to risk management, the use of BIM technology for quality management etc.

We were pleased to host university representatives and industry practitioners who had the opportunity to discuss jointly the proposals regarding the promotional work of our junior colleagues. Thanks to all international participants hoping for further cooperation, we sincerely thank our reliable industrial partners: BASF, Peri, ViaCon and WPIP.

Thanks also to Dr. Marcin Gajzler and Piotr Nowotarski for the perfect organization of the meeting and cooperation with our publisher. I hope that your visit in Poznan also allowed discussions outside the conference room during our organized tours included in social program as well as an individualized tour of our city, which offers many attractions within easy walking distance of the Poznan University of Technology campus (Malta Lake, Old Market Square, Aqua Park Malta etc.).

I would like to thank all the participants and invite all of you with your friends to join our next meetings in the near future – May 2019.

Jerzy Paslawski

edp sciences

Web of Conferences

Statement of Peer review

In submitting conference proceedings to *Web of Conferences*, the editors of the proceedings certify to the Publisher that

1. They adhere to its **Policy on Publishing Integrity** in order to safeguard good scientific practice in publishing.
2. All articles have been subjected to peer review administered by the proceedings editors.
3. Reviews have been conducted by expert referees, who have been requested to provide unbiased and constructive comments aimed, whenever possible, at improving the work.
4. Proceedings editors have taken all reasonable steps to ensure the quality of the materials they publish and their decision to accept or reject a paper for publication has been based only on the merits of the work and the relevance to the journal.

Title, date and place of the conference

3rd International Workshop on flexibility in sustainable construction. April, 20th, Poznan, Poland

Proceedings editor(s):

J.Pasławski
on behalf of International International Programme Committee

G. Girmscheid (Switzerland)
J. Jasiczak (Poland)
A. Kaklauskas (Lithuania)
W. Maliene (United Kingdom)
J. Paslawski (Poland)
M. Radujkovič (Croatia)
A. Sobotka (Poland)
J. Šelih (Slovenia)
Z. Turskis (Lithuania)
L. Ustinovičius (Lithuania)
T. Vilutiene (Lithuania)
T. Mróz (Poland)
J.Tamošaitienė (Lithuania)

Date and editor's signature

12.10.2018

Jerży Pasławski

Methods of construction quality issues elimination with use of modern technology.

Sebastian Dubas[1,*], *Jerzy* Pasławski[2]

[1] Poznan University of Technology, Institute of Structural Engineering, Piotrowo 5 street, 60-965 Poznan, Poland
[2] Poznan University of Technology, Institute of Structural Engineering, Piotrowo 5 street, 60-965 Poznan, Poland

Abstract.
The purpose of this article is to indicate possibility of improving quality in construction thanks to use of new technological solutions. A problem of poor construction defect management, low quality assurance expenses and significant defect costs has been defined. Proposal of quality management tools has been suggested. Article presents set of defects collected throughout construction inspections. Defects were analysed (FMEA), further proposal of quality improvement based on proactive approach with use of modern technology solutions and their estimated influence on risk analysis has been described. The article is summed up with conclusions.

1. Introduction

No one needs to be convinced that construction differs from any other field of economic activity. Many factors indicate that construction differs in relation to the typical production activity in many ways such as: bonding the product to the ground, relatively high unit cost of the product, unitary and often unique production, craft character of production, large direct labour inputs by employees, impact of the changing environment on the processes being implemented, unique composition of the construction project management team such as contracting authority and its inspector, general contractor and subcontractors, suppliers, designers (architect, structural engineer, industry specialists), etc. This specific characteristic of construction generate typical problems in quality management, such as:
• Problems with acquiring qualified staff and employees,
• Frequent conflict situations that enable abuse by shareholders of the construction process,
• Frequent distortions and changes in the conditions of implementation, which significantly changes situation and requires determination of responsibility and indication of restoration path - a return to the normal state,
• Long quality chains that, in the event of problems, make it difficult to determine causes and responsibilities of the parties.

Table 1. Difference between reactive and proactive quality management.

	Reactive	Proactive
1	Problem is identified when in collision with successor works.	The potential problem is identified and checked before proceeding to the next stage.
2	Errors are being verified ad hoc.	Checklists are prepared for staff at the operational level to detect errors at the source.
3	Quality is verified occasionally.	Quality monitoring is a continuous process - in real time.
4	Quality inspections take place on final stages of works.	Quality inspections take place immediately after the completion of every stage of works.

As a result, in the quality management there are enormous opportunities to upgrade and improve construction processes. By analysing the typical stages of quality management development, from quality inspection through quality control, quality assurance, total quality management, we come to Quality 4.0 inspired by the Industry 4.0 concept. Recalling the first industrial revolution based on the introduction of steam, the second introducing new opportunities based on mass production, the third becoming foundation of the digital computer, it is worth pointing out the foundation of the fourth revolution, which are digital-physical systems. Increased use of sensors, mobile devices and wireless communication will enable even better quality improvement thanks to a change of approach from traditional - reactive to new: proactive. The essential features of this change are included in Table 1.

* Corresponding author: sebastian.dubas@put.poznan.pl

Analysing the situation in quality management in various countries, such as Finland (Teravainen et al.) [12], Spain (Prado-Roman et al. 2018) [13], Nigeria (Eze et al.) [14] or the countries of south-eastern Europe (Vukomanovic et al.) [15] various problems specific to individual countries or industry may be identified. However, the proactive approach is undoubtedly a mutual element of quality engineering development in construction in many countries. Departure from the traditional approach may be based on the use of increasingly popular BIM technology (Lin et al.) [16], which seems to be a particularly favourable breakthrough from the point of view of highly digitally undeveloped domain, such as construction. But should companies implement this expensive and demanding well-trained technology staff? Is it possible to improve the quality by using simpler devices and methods before the general use of BIM in construction?

2. Quality problems in construction

Construction industry is struggling with great amount of human labour force and low level of automation and industrialization. The above in combination with a high level of detail and lack of repeatability contributes to generation of a significant number of errors. The education of staff takes a relatively long time in relation to the duration of investment, which is why the employees' experience and proper supervision from managerial staff count on the construction site. In the current state, these two factors affect quality directly, and shortages in any of them result in errors. This is because, no quality control costs are anticipated on planning stage of investment or funds for these activities are relatively low. Lack of budget for supervision is a direct cause of the reactive approach, in which quality is mentioned only when the results of activities do not meet basic requirements and their repair is necessary. Transition to proactive quality management is suggested, which can contribute to the minimization of errors with the help of modern technological solutions, as discussed in this work.

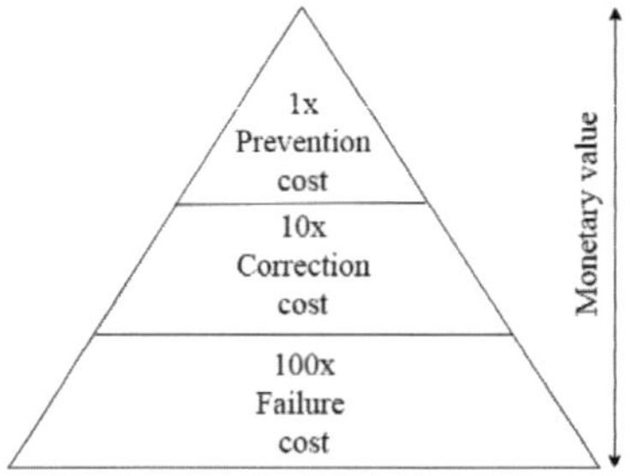

Fig. 1. Lack of quality costs, based on [10].

Severe influence of rework costs on construction companies profit, as revealed by Love et al. [11] (even 28% profit reduction) is a good starting point for proper quality management while planning investment. According to the approach explained by Ross and Perry [10] securing budget for quality assurance may produce income due to savings based on lack of rework and failure costs – it has been shown on monetary estimate example, according to which correction of errors costs 10 times more than their prevention, and what is more cost of failure is 100 times more than preventing it – see fig. 1. Nevertheless defects are hard to anticipate, probability of occurrence is seldom known, impact is unpredictable, that is why it comes easily for decision makers to exclude or cut quality assurance costs form construction budget, because while planning it is hard to combine monetary value of quality management with final outcome of construction.

3. Modern technology for quality management

The computing power of modern computers and increasingly sophisticated software open up new possibilities for construction and quality management. Modern work tools are created, thanks to which design, construction, measurement, continuous and as-built quality control become faster and more accurate. Due to such progress, work expenditures on reactive activities are reduced, thus engineering and managerial staff can devote time to important issues such as proactive way of quality management.

Z. Ma et al. [1] showed collaborative system as a tool for quality management for purpose of on-site inspections. It consists of two main elements: 1) BIM model supplies visual form and digital description of designed elements while 2) indoor positioning system is responsible for locating inspector on construction site and simultaneously inspected elements in BIM model. Tool allows for double staged quality inspection. At first inspection is done by contractors inspector, after positive result of verification, system enables to invite inspector from supervisor for second stage of verification. During each inspection remarks concerning necessary repairs and rework are registered, saved on server and may be distributed among involved parties. System allows for making pictures, selecting target objects, assigning colours to BIM model elements representing whether inspected lot requires second stage of inspection (blue), passed inspection positively (green) or negatively (red). Ma et al. show that system based method of inspection is approximately 50% less time consuming than traditional method.

In work dedicated to defect management with use of augmented reality, Lee et al. [2] present an approach to construction quality inspection with use of image matching technology that compares 3D model and view from camera. Translation visual data taken from camera allows to verify whether visible elements are arranged as supposed to be in designed model for purpose of quality check. What is more, simultaneous view of modeled information and view from camera is recommended for construction workers to allow them to see location, dimensions and size of elements to be built before actual works performance, what results in proactive quality management.
According to El-Saboni [4] research based on 2 case studies and series of interviews, electronic communication systems have positive effects on project in many areas. It is considered that improvement (how significant) occurred in such project success criteria as time (very high), quality (low), minimum variations (moderate), claim management (high), project team satisfaction (very high), master file with lessons learned (high).
Implementation of augmented reality (AR) in construction may be of use in construction quality management. Y. Zhou et al. [5] have studied feasibility of implementing AR into inspection of tunnelling works. Digital measurements have been investigated with use of camera, coordinates transformation program and marker as a base point for global coordinate determination. Inspection needs to be preceded by inspection planning, BIM model preparations and linking to markers.
Terrestial laser scanning (TLS) may be used in surface flatness measurements and may come helpful in quality control after works being performed (for example concrete flooring, earthworks and geodesy) . According to F. Bosché and E. Guenet [6] accuracy of TLS measurements depends on chosen approach towards surface regularity calculation. From chosen methods (Straightedge and F-Numbers) they concluded that both can obtain satisfactory results in quality control, with possible better outcome after applying denser grid setup. What is important, with use of Straightedge approach achieved measurements were more precise and obtained in less time, comparing to traditional manual method.
C. Boton et al. [7] performed case study on 4D BIM simulation in pre-construction and construction phases. Study resulted in showing that 4D simulation in construction phase can be of assistance in coordinating project, managing construction site, works and logistics. Usefulness of information gathered in BIM model depends on Level Of Development (LOD) of design. The higher LOD is helpful in analysing construction details. C. Boton [8] described VR-based 4D simulation preparation process with use of compact and portable simulation system, that may be utilised in on-going construction process.

4. Defects throughout construction

Construction projects characterize diversification of construction schemes, dimensions, spans, support elements, materials, organizational schemes, construction site, plot of ground, location etc. Due to unique and unrepeatable production, each construction faces many risks and is prone to defects. Construction engineers face quality issues and try to prevent defects arising on daily basis. Nevertheless, even with quality management plans and inspections some defects may occur. Case study has been based on list of sample defects collected throughout observations of numerous construction sites and inspections among constructions located in Poznan, Poland and its outskirts. From defect data collected, representative ones have been selected for purpose of further investigation. Defects assembled in Table 2 have diverse background. In presented case diversity of such criterion were juxtaposed:

- area of works during which defect occurred,
- type of defect and its consequence,
- cause and responsible phase of construction stage (design, construction),
- result and losses caused by defects,
- necessary actions required to solve problem,

Analyzing presented set of errors, one can notice basic division into designing and constructing errors. Love et al. [11] refer to them as *project rework* and *construction rework*. The former are mostly associated with processing of a large amount of information, which has to be coordinated in a multi-person team, and then conveyed in a descriptive and graphical form consisting of many interdependent drawings. Construction problems, on the other hand, characterize performance of unique works by construction workers (even within one field of works) in changing environmental conditions, and using a variety of materials characterized by diversified physical properties and applications, behaving differently under the influence of various factors. All of the above components contribute to the difficulties of performing works and are the reason for human errors at every stage of construction production.

Table 2. Defects collation sheet.

No.	Area of works	Description of defect	Cause	Result	Applied solution
1.	Monolith construction	Inapropriate diameter of steel rods in construction elements - walls.	Design: Mistake in drawing. According to calculations, steel rod diameters should be smaller every couple stories, but information included in drawings did not match information included in steel collation sheet.	Loss of material (unnecessary, larger amount of steel than trequired).	Request for information from construction designer. Correction of drawings required.
2.	Monolith construction	Different size of opening in architectural and construction drawings.	Design: Lack of coordination between architectural and construction 2D drawing.	Verification required. Time delay.	Request for information from design team.
3.	Masonry works	Misaligned wall (wrong dimensions of room).	Construction: Human error - wrong measurement.	Time delay. Additional costs of rework. Loss of material (damaged bricks).	Dismantling and rebuilding wall is required.
4.	Material acceptance procedure	During acceptation procedure of a material declared by contractor, some parameters did not match requirements set in design specification.	Construction: Human error - misinterpretation of data in material specification.	Time waste on collecting wrong data. Time delay (due to lack of acceptance works have been delayed).	New material needed to be declared.
5.	Concrete flooring	Exfoliation of top layer of concrete flooring.	Construction: Lack of appropriate concrete curing during high temperatures. Works have not been properly supervised and curing has been forgotten.	Additional costs of rework.	Repair of concrete floor.
6.	Plastering	Mould efflorescence on surface.	Construction: Lack of ventilation while curing. Works have not been properly supervised - humidity in premises rised and liquefaction occured.	Additional costs of rework.	Mould removal. Conservation of surface.
7.	Ground works	Level of ground surrounding building without appropriate slope outwards. Terraced houses level does not correspond to surrounding level of ground.	Design: altitude ordinates not coordinated with actual measurements. Measurements taken for purpose of design were too general. No detailed measurements were provided.	Inappropriate ground levelling. Water flow not arranged. Local buffs. Surrounding fencing subsided and leaned.	Performing wall embankment, ground levelling, fencing corrections.
8.	Ventilation	Vertical ventilation shaft obstructed.	Construction: Concrete from ceiling on above blocked ventilation shaft. Probable cause: no inspection of formworks.	Lack of ventilation in kitchen.	Partial wall dismantling required for concrete removal.
9.	Area measurements	Result of measurements was significantly different (+3%) than the one in design .	Construction: Standard taken for purpose of measurements was different than the one assumed in design.	Falsely buyer has been asked to pay extra costs.	Buyer demanded inspection and area verification, thus mistake has been corrected.
10.	Staircase construction	Level of concrete stairway has been 6cm lowered comparing to design.	Construction: Mistake in measurements - wrong level set during construction.	Steps were too low for purpose of finishing works and lining. Additional costs of rework.	Steps need to be adjusted and lifted.

5. Defects prevention with use of modern technology

The defects shown in point 3. revealed in the course of works performance or inspection carried out, despite the differentiation of the origin areas and the divergent sources of their causes, have one thing in common. With the use of modern technological solutions, it would be possible to avoid their occurrence, reduce their severity or at least detect them faster.
The defect concerning monolithic structure and the incorrect diameters of reinforcing bars described in Table 2 was caused by a design error. As a proposition to avoid creating a mistake, it is possible to develop 3D documentation - BIM model with detailed information allowing such details as reinforcing steel class, diameter and length of relevant items to be included. With use of detailed documentation, table with set of the reinforcing steel would be made automatically and there would be no possibility of error and loss of information during the rewriting.
Another defect related to the incorrect dimensions between architectural and construction drawings was the cause of lack of inter-branch coordination. As in the previous case, when using the 3D BIM documentation, the defect would not have occurred, as the designer of a monolithic structure would have a direct picture of object's dimensions designed by the architect. In a situation in which, however, human error would occur, it could be detected quickly, still at the design stage, by means of the "clash detection" option between the models of individual industries.
The defect of the masonry structure, consisting in improper placement of the masonry wall in relation to the designed dimensions caused by a human error, would be possible to avoid using tools based on the Augmented Reality technology. Thanks to it, the contractor of the masonry structure would be able to visualize location of the wall in relation to the rest of the structure, thanks to which verification of dimensions would take place before the execution. In the event that the technology would not be introduced at the lowest level, but was only available to managers or inspectors, detection of the defect would be faster due to simple, unambiguous evaluation and comparison of reality with the project on common screen during the inspection.
Organizational and formal issues at the construction site can also lead to delays in implementation or defects (incorporation of inappropriate or non-compliant materials etc.). Building documentation contains a lot of data dispersed in many documents (drawing, descriptive), which should be combined in order to extract information about one element. To improve the construction process it would be necessary to merge descriptive documents (technical specification of performance and acceptance of works, technical description, etc.) with design drawings, which is practically achievable thanks to the use of 3D modeling and assigning parameters or references to individual elements. The implementation of detailed BIM project documentation, enriched with precise data on designed elements, materials and requirements, would allow to perform an automated statement of characteristics and a direct comparison of the proposed material in relation to the material being designed. In addition, contractor could easily send material inquiries to its suppliers, directly transferring all data available in the project.

6. Analysis of possible defect prevention

For purpose of establishing discussed defects influence on construction, a failure mode and effect analysis (FMEA) has been carried out twice. Evaluation has been made by 5 representatives of construction (average 9 years of experience). Each interviewee has been asked to evaluate each defect and assign score to an attribute. Presented scores reflect mean value out of all answers. Defects have been ranked according to attributes in given scale:

- Probability [P] – probability of occurrence according to scale: 1 - extremely unlikely, 2 - remote, 3 – occasional, 4 – reasonably possible, 5 – frequent,
- Severity [S] – severity of negative influence according to scale: 1 – no effect, 2 – very minor effect, 3 – minor effect, 4 – critical, 5 – catastrophic,
- Risk Level [RL] – score of multiplication probability by severity [RL]=[P]x[S], RL scale: < 9 – low, < 12 – moderate, < 15 – high, ≥ 16 – unacceptable.

Firstly defects have been evaluated as they are being handled currently on construction site. In second approach they were evaluated as predicted by construction staff, how would they be handled with use of modern solutions (partially explained in point 4 of this elaboration). It is worth noting that according to interviewees severity of defects have not been influenced by introduction of modern technologies. Results of two analysis are presented in table 3.
Performed FMEA analysis shows, that introducing new technologies into construction process, according to construction representatives, could help to diminish probability and increase detection of defects. As a result of proposed actions, risk level in 9 of 10 cases is predicted to be significantly lowered. Only in one case no technological novelty has been proposed. Due to specificity of ventilation obstruction defect, it would be hard to detect it with use of known visual aided tools because of superficial invisibility and inaccessibility of fault. Measurements of air flow or placing inspection camera in ventilation shaft would result in defect detection, which is how this defect has been detected in the first place. Nevertheless, as a result of interview, when asked about implementation of these measures into construction site, each representative noticed high cost of introduction as a barrier.

Table 3. FMEA of introducing modern technology solutions for defect prevention.

No.	Area of works	Description of defect	First risk analysis				Introduced solution	Second risk analysis			
			[P]	[S]	[P]x[S]	Risk Level		[P]	[S]	[P]x[S]	Risk Level
1.	Monolith construction	Inapropriate diamterer of steel rods in construction walls.	4	3	12	High	BIM modelling with construction parameters	1	3	3	Low
2.	Monolith construction	Different size of opening in architectural and construction drawings.	4	5	20	Unacca ptable	BIM modelling with branch coordination	1	5	5	Low
3.	Masonry works	Misaligned wall (wrong dimensions of room).	3	4	12	High	Augmented Reality	2	4	8	Low
4.	Material acceptance procedure	During acceptation procedure of a material declared by contractor, some parameters did not match requirements.	4	5	20	Unacca ptable	BIM modelling with material parameters	2	5	10	Moderate
5.	Concrete flooring	Exfoliation of top layer of concrete flooring.	4	5	20	Unacca ptable	4D BIM interactive schedule	2	5	10	Moderate
6.	Plastering	Mould efflorescence on surface.	5	3	15	Unacca ptable	4D BIM interactive schedule	4	3	12	High
7.	Ground works	Level of ground surrounding building without appropriate slope outwards.	2	5	10	Modera te	BIM modelling with scanning input data	1	5	5	Low
8.	Ventilation	Vertical ventilation shaft obstructed.	3	5	15	Unacca ptable	No modern solution proposed.	3	5	15	Unaccap table
9.	Area measure-ments	Result of measurements was significantly different (+3%) than the one in design.	3	4	12	High	BIM and scanning	2	4	8	Low
10.	Staircase construction	Level of concrete stairway has been 6cm lowered comparing to design.	4	4	16	Unacca ptable	Augmented Reality	3	4	12	High

7. Summary and conclusions

Until now, none of the developers on the Wielkopolska market has kept a digital record of defects [9], let alone comprehensive quality management or implementation of investments in BIM. Considering the planned late start of the implementation of BIM technology in Poland (the implementation horizon is estimated to be in 10 years), the question arises whether it remains only to accept errors in construction projects until then.

Analysis of the current situation in construction in our marketing environment, analysis of the case study and possible solutions, allows to draw the following conclusions:

1. The most common cause of defects in construction is human error. This is the cause of many disadvantages which, in the face of complex implementation and design conditions, are difficult to avoid in the current form of conducting construction investments.
2. Low quality management expenditures cause a large number of re-processing costs and delays.
3. The inclusion of the FMEA method adapted to specifics of construction production may be the basis for systematic improvement of quality in construction processes.
4. Modern technologies are an excellent basis for improving the quality of construction projects, but the high cost of their implementation is a serious barrier to entry for both design and construction companies.

The authors see the possibility of a quick implementation of a simple and economically justified system that will streamline selected construction processes and support controlling activities to minimize defects and, above all, prevent them (proactive approach). The proposed solution can serve as a relatively quick improvement of current situation, and serve as a bridge between the current state of affairs and the future of the digital structure, which is the subject of further research.

Acknowledgements

The authors are grateful for supporting the research described in this paper from the statutory funds of the Institute of Structural Engineering of Poznan University of Technology 01/11/DSPB/0003.

References

1. Z. Ma, S. Cai, N. Mao, Q. Yang, J. Feng, P. Wang, Automat. Constr., *Construction quality management based on a collaborative system using BIM and indoor positioning*, **92,** 35–45 (2018)
2. J.Y. Lee, J.So. Choi, O.S. Kwon, C.S. Park, 2012 International Conference on Information Science and Applications, *A Study on Construction Defect Management Using Augmented Reality Technology*, May 2012
3. A. Mills, P. E. D. Love, P. Williams, Journal of Construction Engineering and Management, *Defect Costs in Residential Construction*, January 2009
4. M. El-Saboni, G. Aouad, A. Sabouni, Advanced Engineering Informatics, *Electronic communication systems effects on the success of construction projects in United Arab Emirates*, **23**, 130-138 (2009).
5. Y. Zhou, H. Luo, Y. Yang, Automat. Constr., *Implementation of augmented reality for segment displacementinspection during tunneling construction*, **82**, 112-121 (2017).
6. F. Bosché, E. Guenet, Automat. Constr., *Automating surface flatness control using terrestrial laser scanning and building information models*, **44**, 212–226 (2014).
7. C. Boton, S. Kubicki, G. Halin, Procedia Engineering, *The Challenge of Level of Development in 4D/BIM Simulation Across AEC Project Lifecyle. A Case Study*, **123**, 59-67 (2015).
8. C. Boton, Automat. Constr., *Supporting constructability analysis meetings with Immersive Virtual Reality-based collaborative BIM 4D simulation*, **96**, 1-15 (2018).
9. S. Dubas, J. Pasławski, Procedia Engineering, *The concept of improving communication in BIM during transfer to operation phase on the Polish market*, **208**, 14-19 (2017).
10. J. E. Ross, S. Perry, Total Quality Management, CRC Press, (1999).
11. P. E. D. Love, Z. Irani, F. Ackermann, P. Teo, Production Planning and control, *The cost of Rework : Insights from Construction and Opportunities for Learning* (2018).
12. V. Teravainen, J.M. Junnonen, S. Ali-Loytty, Construction Economics and Building, *Organizational culture: Case of the Finnish construction industry*, **18**, 48-69, (2018).
13. C. Prado-Román, C. Mercado-Idoeta, C. Del Castillo Peces, J. Del Castillo Peces, Cuadernos de Gestion, *The effects of implementing ISO 9001 in the Spanish construction industry*, **18**, 149-172, (2018).
14. E. C. Eze, J. E. Idiake, B. O. Ganiyu, Independent Journal of Management & Production, *Rework risks triggers in the nigerian construction industry: a view of built environment professionals*, **9**, (2018).
15. M. Vukomanović, M.M. Nahod, M. Radujkovic, Journal of Civil Engineering and Management, *EFQM excellence model as the TQM model of the construction industry of southeastern Europe*, **20**, 70-81, (2014).
16. Y. Ham, K.K. Ham, J. J. Lin, M. Golparvar-Fard, Visualization in Engineering, *Visual monitoring of civil infrastructure systems via camera-equipped Unmanned Aerial Vehicles (UAVs): a review of related works*, **4**, (2016).

Impact of the adopted strategy on the result of multi-criteria analysis of technology solution based on AHP (BOCR)

Agnieszka Dziadosz[1*], *Dominika* Banach[1], *Wiesław* Meszek[1] and *Mariusz* Rejment[2]

[1]Poznan University of Technology, Piotrowo 5 Street, 60-965 Poznan, Poland
[2]Wroclaw University of Science and Technology, Wybrzeże Wyspiańskiego 27, 50-370 Wroclaw, Poland

Abstract. The complexity of problems and the necessity to find a comprehensive approach to their solution favours the search for various methods that support the decision making process in the construction industry. At present, the multi-criteria methods lead the way. They enable the assessment of the problem in many aspects and with taking into account qualitative and quantitative data. Moreover, these methods do not limit the decision-maker in terms of the number of considered variants as well as the scope and the type of adopted assessment criteria. The aim of the article is to present the procedure of assessment of proposed building material solutions with taking into account the division of criteria due to their character – benefits (B) and costs (C). It was shown the dependence adopted by the decision maker strategy on the final ranking of solutions using the BOCR approach in the AHP method.

1 Introduction

The multi-criteria decision making methods are very useful support for solving complex problems. These problems often have to be assessed in a multi-aspect and comprehensive manner. It is not only important to identify the problem. It is also important to diagnose the proposed solutions by the decision maker for a given problem. The literature on the subject cites numerous examples of the use of the aforementioned methods, at the same time presenting their diversity, the procedure of aggregation of results, the possibility of combining by creating so-called "hybrids" [1–8]. These methods have many advantages, and one of the few disadvantages is the subjective judgment made by the decision maker. In the case of solving problems, in which we must take into account both qualitative and quantitative criteria, the basis for making the decision is the approximate result of the analysis and the initial ranking of variants. The authors of the paper checked the degree of dependence of the final ranking of variants of brickwork of the external, structural wall

* Corresponding author: agnieszka.dziadosz@put.poznan.pl

from the strategy adopted by the decision maker, i.e. minimizing of costs or maximizing of profits based on the AHP method.

2 AHP method and its BOCR extension

This article presents a procedure based on the method of the Analytic Hierarchy Process (AHP), which takes into account the division of criteria into favourable and unfavourable from the decision-maker's point of view. The AHP method, developed by Thomas L. Saaty in the 1970s, is to rank the considered variants from the best to the worst [9]. The author of the AHP method introduced a 9-point scale for pairwise comparison, wherein the value of 1 means the equivalence of two criteria (factors, alternatives, etc.), and 9 means the absolute advantage of one criterion over the other. To facilitate the calculations, the considered pairs are set together in the matrix, comparing successively the elements from the first column with the elements of the top row.

However, the authors of the following works [10, 11] proposed the division of criteria by their character as well as by taken into account in the procedure of the criteria selection the criteria that are opposite to others. The criteria are ranked in 4 groups, namely: benefits (B), opportunities (O), costs (C) and risks (R). For each of these groups, the rank is also assigned according to the AHP idea. Then, after performing the relevant calculations, the partial results are aggregated into the totality, i.e. in the final assessment of the construction investment in accordance with the method of results aggregation:

bB + oO – Cc – rR,

where:

b, o, c, r – are weights of groups,

while:

B, O, C, R – are the data obtained from the analysis for each considered alternative obtained as a result of the AHP comparison.

We have 4 possibilities of combining the obtained results: B, O, C, R with appropriate normalized weights: b, o, c, r, namely:

– bB + oO + c (1/C) + r (1/R),
– bB + oO + c (1 – C) + r (1 – R),
– bB + oO – cC – rR.

However, the BO/CR ratio (mentioned as the fourth option) does not take into account the normalized weights: b, o, c, r.

The procedure adopted for preparing variant of solutions:

1. Preparing of matrix of comparison in pairs for criteria in order to determine their mutual validity and for alternatives to a given criterion.
2. Pairwase comparison of alternatives with individual criteria according to the AHP idea.
3. Calculations of the vector of weights importance.
4. Calculation of λ_{max} (the maximum eigenvalue of the matrix).
5. Calculation of the consistency index ($CI = \lambda_{max} - n / n - 1$).
6. Choosing of the right value of the random consistency ratio (RI).
7. Calculation of the consistency ratio (CR = CI/RI).
8. Checking of the consistency of the pairwise comparison matrix (if the comparisons in the matrix are consistent or not; the inconsistency of the matrix is allowed to be – 0,1 CR <0,1).
9. Aggregation of results according to the first or the second method:

$$b * B + o * O + c * \frac{1}{C} + r * \frac{1}{R} \quad (1)$$

$$b * B + o * O + c * (1 - C) + r * (1 - R) \quad (2)$$

3 Case study – construction layer of wall

1.1 Proposed variants of building material solutions

Analyzing the design documentation, the authors have proposed 3 variants of making the construction layer of a 3-layered exterior wall for a single-family building (Table 1). The quantity of brickwork (calculated in the bill of quantities) for the construction layer was 204,27m2.

Table 1. Analyzed variants – building material

Symbol	Variant
P	Ceramic block Porotherm 25 Dryfix
Y	Aerated concrete block Ytong Forte PP 2,5/0,4
S	Silicate block Silka E 24

1.2 Estimation criteria

In the presented example, the authors have done the building material solutions assessment according to the classical AHP approach and after considering the BOCR development due to two groups of criteria: benefits – B and costs – C (Table 2 and Table 3). The application of the AHP method was dictated by its numerous advantages, i.e.:
• arrangement of factors, criteria and variants,
• the possibility of using quantitative and qualitative criteria,
• easy to understand course of proceedings,
• flexible model for a wide range of multi-criteria problems,
• no need to use specialized software,
• application of many criteria in order to make decision – multi-dimensionality of the problem.

Table 2. Selection criteria for Benefit (B)

Symbol	Criteria
KB1	Overall heat transfer coefficient – U [W/m2K]
KB2	Sound insulation
KB3	Compressive strength

Table 3. Selection criteria for Costs (C)

Symbol	Criteria
KC1	Building material purchase cost

KC2	Transport costs
KC3	Labour cost

1.3 Evaluation and analysis of results

The assessment of variants for benefits and costs is presented in Figure 1 and Figure 2 as well as in Table 4 without taking into account weights for the adopted strategies: maximizing of profit or minimizing of costs.

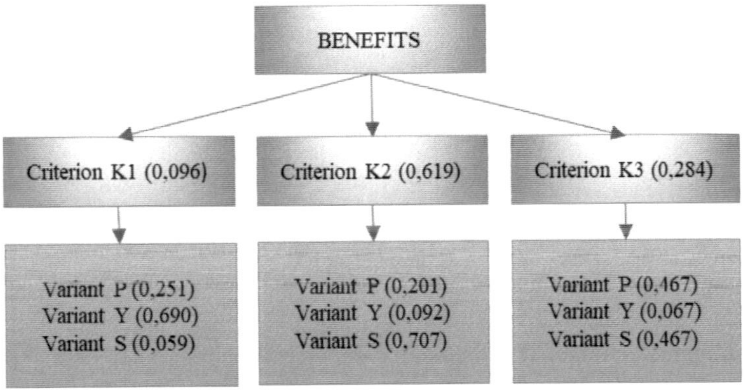

Fig. 1 Summary of results for benefits

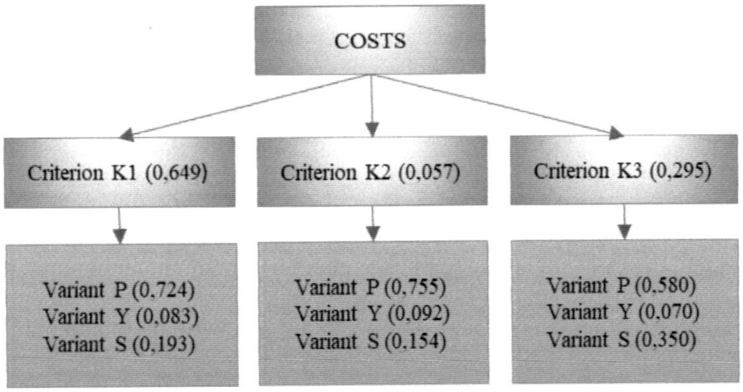

Fig. 2. Summary of results for costs

Table 4. Summary of results for benefits and costs

Building material	Benefits	Costs
P	0,282	0,683
Y	0,142	0,080
S	0,576	0,237

For the construction layer, the optimal solution, from the benefit perspective, was the variant of the wall made of silicate blocks – Silka, and from the cost perspective it was a variant of ceramic blocks – Porotherm. The introduction of the scenarios with weights has significantly changed the ranking. For the first equation, the best solution was made of aerated concrete blocks – Ytong, for both benefits and costs. For the second equation, due to the benefits, the optimal variant was silicate blocks – Silka, due to the costs – aerated concrete blocks Ytong (Tables 6, 7 and Figures 3, 4). In order to aggregate the results of the BOCR analysis, a scenario for the values of weights b and c was prepared (Table 5).

Table 5. Scenarios of weight values: b and c

	Scenarios								
	9/1	7/1	5/1	3/1	1/1	1/3	1/5	1/7	1/9
b	0,900	0,875	0,833	0,750	0,500	0,250	0,167	0,125	0,100
c	0,100	0,125	0,167	0,250	0,500	0,750	0,833	0,875	0,900

Table 6. Analysis of results depending on the adopted weight scenario for the first equation

	Scenarios								
	9/1	7/1	5/1	3/1	1/1	1/3	1/5	1/7	1/9
P	0,400	0,429	0,479	0,577	0,873	1,168	1,267	1,316	1,346
Y	1,378	1,687	2,207	3,233	6,323	9,413	10,439	10,959	11,268
S	0,941	1,032	1,185	1,487	2,398	3,308	3,611	3,764	3,855

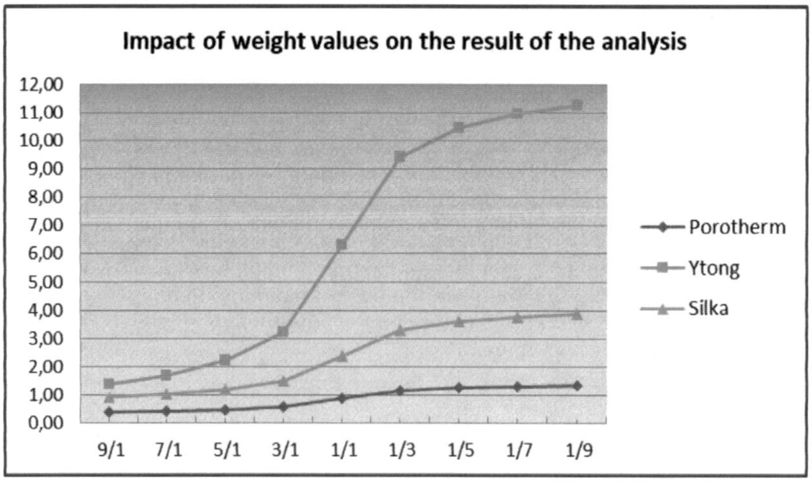

Fig. 3. The influence of weight values on the result of the analysis for the first equation

Analyzing the impact of weight values on the final result, it can be seen that in the case when for the decision-maker, the benefits are more important, the differences between options are convergent and there is not much variation between the alternatives from the

benefits perspective. The optimal solution would be aerated concrete blocks Ytong. Other variants are classified lower in the chart than aerated concrete blocks Ytong.
If, however, the interest is increased from the cost perspective, the discrepancy between the variants is significant and then the best solution would be a variant with aerated concrete blocks Ytong. The other two variants slightly increase compared to the beginning.

Table 7. Analysis of results depending on the adopted weight scenario for the second equation

	Scenarios								
	9/1	7/1	5/1	3/1	1/1	1/3	1/5	1/7	1/9
P	0,285	0,286	0,287	0,290	0,299	0,308	0,311	0,313	0,313
Y	0,220	0,239	0,272	0,337	0,531	0,726	0,790	0,823	0,842
S	0,595	0,600	0,607	0,623	0,670	0,716	0,732	0,740	0,744

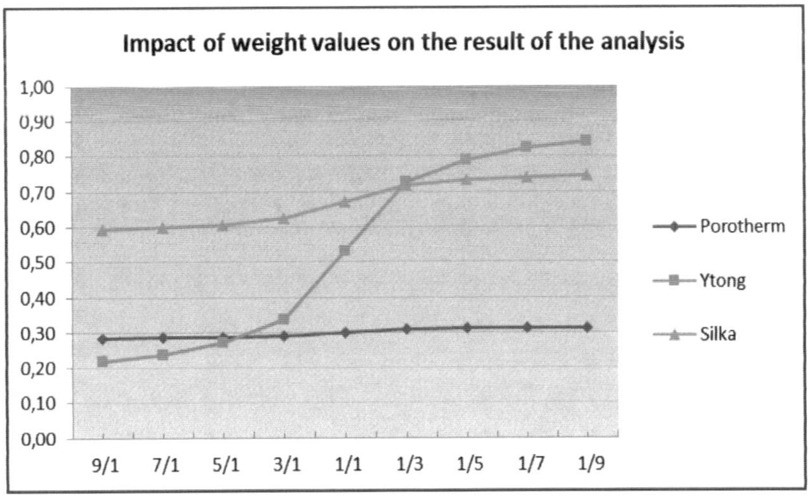

Fig. 4. The influence of weight values on the result of the analysis for the second equation

Analyzing the impact of weight values on the final result, it can be seen that in the case when for the decision-maker the benefits are more important, the differences between the variants are large, then silicate blocks Silka would be an optimal solution. Values for alternatives with ceramic blocks Porotherm and aerated concrete blocks Ytong are convergent.

If, however, the interest is increased from the cost perspective, the discrepancy between the variants varies dramatically and then the variant with aerated concrete blocks Ytong would be the best solution. However, the similar values are also occurring for the variant with silicate blocks Silka.

4 Conclusions

In conclusion, a comprehensive assessment of the considered alternatives allows the decision-maker to minimize the risk of making the wrong decision. The use of multi-criteria methods allows avoiding randomness in the decision-making process. It also allows taking into account the requirements and preferences of the decision maker and allows to consider the problem in the multi-dimensional way. Using the development of the AHP-BOCR method, it is possible to comprehensively analyze of variants with the isolation of those factors that are most important for the decision maker. It is worth emphasizing that the final result of the selection is influenced by the strategy adopted by a decision maker: maximizing of profit or minimizing of costs. As the presented example shows, the final ranking depends on the adopted weights for the assumed strategies, as well as depends on the method of the results aggregation.

The question of subjectivity of ratings remains with the use of multi-criteria methods. We need an approximate result more and more often, which is the premise for making a decision and a preliminary ranking of variants for solving the problem rather than giving the exact value.

References

1. R. Kozik, AIP Conf. Proc. **1863**,1, 230016, ICNAAM 2016 (2017)
2. A.Dziadosz, Tech.Trans., (1-B), 41-51 (2008)
3. Gajzler M. A.Dziadosz, P.Szymański, Tech. Trans. (1-B), **2**, 107, 71-84 (2010)
4. A.Dziadosz, M.Rejment (1-B), **5**, 221-228 (2014)
5. R.Kozik, AIP Conf. Proc., ICNAAM 2015 (2016)
6. A. Dziadosz, Proc. 5th Nordic Conf. on Constr. Econ.and Organ., **1**, 213-222 (2010)
7. H. Hashemi, S. M. Mousavi, E. K. Zavadskas, A. Chalekaee, Z.Turskis, Sustain. 10 (**5**), 1635 (2018)
8. M. Gajzler, K.Zima, Inter. J. of Civil Eng., 15 (**5**), 641–52 (2017)
9. T.L. Saaty, J. of Sys. Scien. and Sys. Eng., 13, **1**, 1-35 (2004)
10. T.L.Saaty, Y. Choy , Soc.-Econ.Plan. Scien., 35, 243-252 (2001)
11. T.L. Saaty, M. Ozdemir, Math. Comp. Model., 37, 1063-1075 (2003)

Contactless approach to determine pavement skid resistance for Pavement Management System

Jakub Fengier[1,*], *Mieczysław* Słowik[1] , and *Andrzej* Pożarycki[1]

[1]Poznan University of Technology, Institute of Civil Engineering, Piotrowo 5, 60-965 Poznań, Poland

Keywords: Pavement Management System, skid resistance, pavement surface

Abstract. Standard method to assess the pavement profile is to calculate the MPD (Mean Profile Depth) index based on results obtained usually by usage of laser techniques. In analysis the models of the surface pavements have been used in order to calculate the s_1, s_2, s_3 parameters values corresponding to mega-, macro- and microtexture respectively. The values of the developed parameters s_1, s_2, s_3 are calculated from the specific power spectral density values of surface roughness obtained for the threshold pavement roughness wavelength equal to 0.1, 0.05, 0.005 and 0.0005 m. The skid resistance has been correlated to the s_1, s_2, s_3 parameters using 11 varied cases related to asphalt and concrete pavements. Skid resistance tests have been performed using CSR (Continuous Skid Resistance) device with fixed slip ratio equal to 13%. Three different test speed values 45, 65 and 95 km/h have been used. The obtained results lead to factorial correlation equations between developed parameters and skid resistance indices. Correlation results for uncontaminated pavement surface can be characterized by the coefficient of determination values in range between 0.55 and 0.94. The results can be used for contactless determination of pavement skid resistance in Pavement Management System.

1 Introduction

Road pavement diagnosis is a part of Pavement Management System (PMS). To manage the road network there is need to assess each parameter in reliable and effective way. In this paper there has been performed an attempt to evaluate skid resistance by using the model of pavement surface roughness. Laser based parameters are widely used to assess the pavement surface profile. Examples of the measurement methods are LTS [1], CTMeter [2] or SUPPORT system [3]. Other method is to measure force needed to drag the rubber elements on the pavement surface. In this case DFT [2] and British Pendulum Tester [4] should be mentioned. Results of measurement consisting of both methods have been correlated with (commonly used on Polish national roads) skid resistance tester - SRT [5].

* Corresponding author: jakub.fengier@put.poznan.pl

Another step is to evaluate the pavement surface. The first attempt was based on laser method to obtain the shape of surface of the pavement [6]. Photogrammetric method is a relatively new idea to assess the surface of the pavement. This method is based on the few high resolution photos of the same pavement area from different perspectives. The solution is shown in papers [7], [8] and first Polish approach is described in paper [9].

Pavement skid resistance research is not limited to looking for the highly correlated friction coefficients with pavement surface parameters, but also there are first attempts to solve this issue in an analytical way. Among the most relevant approaches is the paper [10] where author formed the equation between friction coefficient and surfaces roughness of pavement and tire. Another attempt worth mentioning is empirical attempt to use that equation in road pavement measurements [11, 12]. Other approach is described in paper [13] where author assumes that skid resistance mostly depends on upper part of pavement surface.

The attempt of correlation between parameters based on pavement surface and skid resistance index is shown in this paper.

2 Methods

2.1 Skid resistance test

Skid resistance was tested using CSR (Continuous Skid Resistance) device shown in fig. 1. CSR is a device that allows to extract from the measurements the both vertical force values taking effect on measurement wheel and torque values. The slip ratio used in the device is equal to 13%.

The skid resistance index based on CSR device is determined by the formula 1:

Equations should be centred and should be numbered with the number on the right-hand side.

$$CSRi = M/(F \cdot r) \quad (1)$$

where: M – torque values measured by using the moment transducer [Nm]
F – vertical force [N]
r – distance between pavement surface and axis of the measuring wheel [m].
More details about CSR device provided in paper [9]

2.2 Parameters based on pavement surface model

Parameters called s_1, s_2 and s_3 have been introduced to enable the classification of the roughness of the pavement surface in general. Most commonly used classification of the roughness of the pavement surface includes [14]:
-microtexture (wavelength lesser than 0.5 mm and amplitude lesser than 0.2 mm)
-macrotexture (wavelength between 0.5 and 50 mm, amplitude between 0.1 and 20 mm)
-megatexture (wavelength between than 50 and 500 mm, amplitude between 0.1 and 50 mm)

All three parameters affect on skid resistance; that is the reason to perform model consisting of all three of them.

Calculation of the s_1, s_2, s_3 parameters is a process which contains the following steps: 1) taking the high resolution photos using the method described in [9], 2) developing the model of the pavement surface (calculation of co-ordinates of points forming the discrete surface) [9], 3) calculation of power spectral density for 2D model of achieved surface, 4) determine the value of power spectral density for specific wavelengths and 5) calculating the values of s_1, s_2, s_3 parameters.
Calculation of the power spectral density surface roughness was based on the co-ordinates obtained from photogrammetric analysis by using the following equation, adopted after [15].

$$Cq= (\Delta x \cdot \Delta y)/(m \cdot n[2\pi]^2) \cdot |FFT^{2D}|^2 \quad (2)$$

where:
Cq is the two-dimensional surface roughness power spectrum. Cq values calculated in two directions are determined in wavenumber domain. Cq consist of values corresponding to each wavenumber of surface roughness present on the observed area; Δx and Δy are the lateral resolution of co-ordinates of model of the pavement within a grid of $m \times n$ data points in the x- and y-directions, respectively; $|FFT^{2D}|$ – absolute value from 2D Fast Fourier Transform,
From calculated Cq values corresponding to the wavelengths of the pavement roughness: 0.0005 m, 0.005 m, 0.05 and 0.1 m which are related to wavenumbers 2000 m^{-1}, 200 m^{-1},20 m^{-1} and 10 m^{-1}, respectively. Parameters s_1, s_2 and s_3 have been calculated from following equation:

$$s_i=(Cq(a)-Cq(b))/(a-b) \quad (3)$$

where: a and b is wave number (for s_1 a=20 m^{-1}, b=10 m^{-1} for s_2 a=200 m^{-1}, b=20 m^{-1} for s_3 a=2000 m^{-1}, b =200 m^{-1})

3 Experimental program

In order to create a pavement surface model and determine skid resistance from its features, the pavement skid resistance and surface roughness tests have been made. The test sections derived from both asphalt pavements (uncontaminated and contaminated) and concrete pavement (with and without marking). For evaluate the pavements surfaces, several photogrammetric points have been performed on each section. The test has been performed on road and airport pavements. The extracted test sections of 100 m length were considered. As far as the CSR device is a continues-like measurement device, the CSRi values have been reported for every subsection. In every section several repetitions have been performed (due to state of the section and availability).The skid resistance tests have been performed using three test velocities 45, 65, 95 km/h. In this paper each section has been marked as:

- nov (section localized on taxiway named November). SMA wearing course was paved 2 weeks before tests, 6 repetitions have been performed.
- h4 (section localized on taxiway named Hotel 4). SMA wearing course was paved 2 weeks before tests, 10 repetitions have been performed.
- h5 (section localized on taxiway named Hotel 5), SMA wearing course was paved 2 weeks before tests, 10 repetitions have been performed.
- RWc (section localized on runway pavement without contaminations). Concrete surface course was paved 2 years before tests. Aggregate used in concrete was granite crushed stone 0/22. Thickness of the concrete layer was 30 cm, concrete class was C35/45 and exposure class was XF4. Brushing process was used to texture the surface of the layer.

Few weeks before tests the hydrophobization process was done. 6 repetitions have been preformed.

- RWr section localized on runway contaminated by rubber, 6 repetitions have been performed.
- RWm section localized on runway with marking, 6 repetitions have been performed.
- E (section localized on Echo 1 taxiway), cracked concrete wearing course with 0/22 aggregate. This taxiway is planned to be repaired due to its technical condition, 3 repetitions have been performed.
- Gc (section localized on taxiway excluded from traffic), concrete wearing course paved decades before tests, 3 repetitions have been performed.
- Gm (section localized on taxiway excluded from traffic with freshly made marking), concrete wearing course paved decades before tests, 3 repetitions have been performed.
- pc section on selected street in Poznan. Asphalt wearing course was paved 2 years before tests (test section after cleaning process), 3 repetitions have been performed.
- pb section on street mentioned above. Asphalt wearing course was paved 2 years before tests (test section before cleaning process, contaminated with mineral contamination from nearby parking entrance with ground pavement), 3 repetitions have been performed.

Number of each cases was 109 and after using the statistical Grubb's test [16] used to detect outliers to further analysis 96 cases were taken into account. The same result has been obtained using 3 sigma rule [16].

4 Analysis of the results

The results of s_1, s_2 and s_3 parameter obtained on each section in function of CSRi calculated for that section has been shown in figure 1.
Three types of regression equation were used in order to determine the relation between s_1, s_2 and s_3 values and the skid resistance index CSRi:
Linear equation

$$CSRi\ (v)=w+a_1\ s_1+a_2\ s_2+a_3\ s_3 \tag{4}$$

Factorial equation

$$CSRi(v)=w+b_1\ s_1+b_2\ s_2+b_3\ s_3+b_4\ s_1\ s_2+b_5\ s_1\ s_3+b_6\ s_2\ s_3+b_7\ s_1\ s_2\ s_3 \tag{5}$$

2nd degree polynomial (quadratic) equation

$$CSRi(v)=w+c_1\ s_1+c_2 s_1^2+c_3\ s_2+c_4\ s_2^2+c_5\ s_3+c_6\ s_3^2 \tag{6}$$

Where: w – CSRi intercept
$a_i/b_i/c_i$ – regression coefficients
v – test velocity

4 Discussion

Modeling the skid resistance of concrete and asphalt pavements without contamination using s_1, s_2 and s_3 parameters allows to achieve the coefficient of determination values r^2 0.85 to 0.90 and 0.55 to 0.70, respectively. The largest values of r^2 were obtained when one used the factorial equation. It was found that they need the largest number of regression coefficients to be determined. According to the generally accepted trend, the CSRi intercept

in this model decreases when the test speed increases. There were no unambiguous rends for the other values of this model along with the increase of the test speed.
The residual values analysis leads to the assumption that following factorial equations (7, 8, 9, 10, 11, 12) are characterized by distribution of residual values similar to normal distribution. A part of other equations residual values distribution differ significantly from normal distribution. Some of the variables were statistically insignificant due to $\alpha=0.05$, presented equations have all factors including statistically insignificant. No tendency of particular variable has been noticed due to statistical significance.
When combining datasets from asphalt and concrete pavements without contamination regression equations lead to decrease (comparing to results from separated datasets) coefficient of determination r^2 from 0.02 to 0.15.
The combination of datasets for modeling on concrete surfaces with rubber contamination and marking levels lead to reducing the r^2 to range of values between 0.04 and 0.30.
However, the photogrammetric methods lead to successful determination of the pavement skid resistance and the most successful models are shown (equations 7-12):
for asphalt pavements without contaminations:

$$CSRi=0.61-3.62E15\cdot s_1-4.46E16\cdot s_2-3.13E20\cdot s_3+3.24E33\cdot s_1\cdot s_2+5.65E33\cdot s_1\cdot s_3+2.76E38\cdot s_2\cdot s_3-5.18E54\cdot s_1\cdot s_2\cdot s_3 \quad (7)$$

$$v=45km/h;\ r^2=0.55$$

$$CSRi=0.58-4.22E15\cdot s_1+1.43E16\cdot s_2-3.73E20\cdot s_3+2.90E33\cdot s_1\cdot s_2+6.86E36\cdot s_1\cdot s_3+2.72E38\cdot s_2\cdot s_3-5.40E54 s_1\cdot s_2\cdot s_3 \quad (8)$$

$$v=65km/h;\ r^2=0.60$$

$$CSRi=0.57-3.57E15\cdot s_1+1.34E16\cdot s_2-3.67E20\cdot s_3+4.37E33\cdot s_1\cdot s_2+6.28E36\cdot s_1\cdot s_3+3.63E38\cdot s_2\cdot s_3-6.32E54\cdot s_1\cdot s_2\cdot s_3 \quad (9)$$

$$v=95km/h;\ r^2=0.70$$

for concrete pavements without contaminations:

$$CSRi=0.75-2.42E17\cdot s_1-1.30E19\cdot s_2-3.72E20\cdot s_3+9.25E36\cdot s_1\cdot s_2+3.11E38\cdot s_1\cdot s_3+5.51E40\cdot s_2\cdot s_3-3.37E58\cdot s_1\cdot s_2\cdot s_3 \quad (10)$$

$$v=45km/h; r^2=0.94$$

$$CSRi=0{,}73-3.00E17\cdot s_1-1{,}23E19\cdot s_2-2.71E21\cdot s_3+9.64E36\cdot s_1\cdot s_2+2.28E39 s_1 s_3+7.99E40 s_2 s_3-5.98E58\cdot s_1\cdot s_2\cdot s_3 \quad (11)$$

$$v=65km/h;\ r^2=0.88$$

$$CSRi=0.74-2.81E17\cdot s_1-1.17E19\cdot s_2-3.35E21\cdot s_3+9.25E36\cdot s_1\cdot s_2+2.23E39\cdot s_1\cdot s_3+8.19E40 s_2 s_3-5.66E58\cdot s_1\cdot s_2\cdot s_3 \quad (12)$$

$$v=95km/h;\ r^2=0.85$$

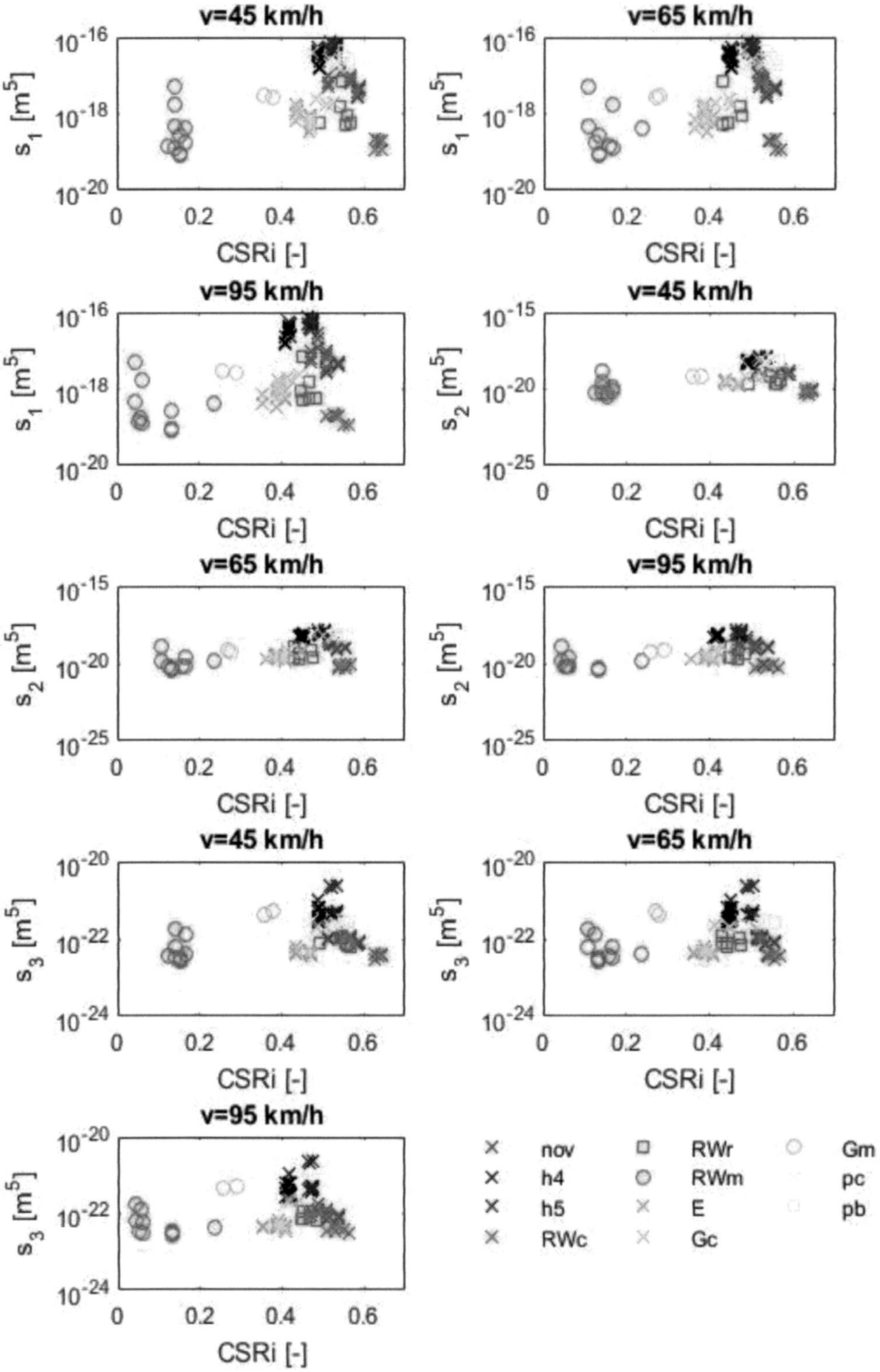

Fig. 1 Results of calculated s1, s2 and s3 parameters in CSRi function for different test velocity

Datasets including contaminated and uncontaminated pavement surfaces require to use two different approaches. The most successful models are related to uncontaminated pavements. No unambiguous relation between s_1, s_2, s_3 and CSRi(v) can be performed using linear equations. None of the considered parameters affect significantly skid resistance on its own.

4 Conclusions

Based on the results of the skid resistance tests and pavement roughness obtained from photogrammetric method the relationship between these values was specified. Three different types of equations were used in order to calculate skid resistance index from values based on the power spectral density of pavement roughness.

Taking into account the contaminated pavement surfaces, significant different regression equations were discovered. In general, the coefficient of determination values are lower for results obtained from the measurements made on contaminated pavements comparing to values achieved from equations based on results obtained on uncontaminated pavement surfaces. However, the most successful models are related to measurements made on pavements without contaminations.

In order to estimate more reliable function describing skid resistance parameters there is a possibility to specify macrotexture by MTD (stand for texture depth) and MPD (stand for profile depth) and microtexture corresponding to BPT (British Portable Tester). Consideration of those parameters and proposed testing method of determination s_1, s_2, s_3 can lead to reliable model used to calculate skid resistance parameters in contactless way. Further studies can lead to the tool which can be use in Pavement Management System.

References

1 P. Vecura, L. Lane, D. Cheng, Caltrans Using Fog and Rejuvenating Seals on State Highways for Preventative Maintenance, (2013)

2 M. Jackson, Harmonization of Texture and Skid-Resistance Measurement Florida Department of Transportation, (2008)

3 I. Wyczałek, J. Stróżyk-Weiss, M. Wyczałek, Archives of Photogrammetry, Cartography and Remote Sensing, **28,** (2016)

4 I. Asi, Building and Environment, **42,** (2007)

5 M. Wasilewska, W. Gardziejczyk, P Gierasimiuk, Transportation Research Procedia, **14**, (2016)

6 Q. Liu, L. Kavanagh, A. Shalaby, B. Izevbekhai, Transportation Research Record: Journal of the Transportation Research Board, **2591**, (2015)

7 A. Slimane, M. Khoudeir, J. Brochard, M. Do, Wear, **264**, (2008)

8 V. Knyaz V, A. Chibunichev, The International Archives of the Photogrammetry, Remote Sensing and Spatial Information Sciences, **XLI-B5**, (2016)

9 A. Pożarycki , J. Fengier , M. Wyczałek , P. Skrzypczak, M. Wesołowski, K. Blacha, Road engineering, **3**, (2017)

10 B. Persson, Journal of Physics: Condensed Matter, **18**, (2006)

11 A. Ueckermann, D. Wang, M. Oeser, B. Steinauer, Journal of Traffic and Transportation Engineering, **2**, (2015)

12 A. Ueckermann, D. Wang, M. Oeser, B. Steinauer, International Journal of Pavement Engineering, **16**, (2015)

13 M. Kanafi, A. Tuononen, Tribology International, **107**, (2017)
14 P. Gierasimiuk, W. Gardziejczyk, Rzeszow University of Technology Research Notebooks, **59**, (2012)
15 M. Kanafi, A. Kouosmanen, T. Pellinen, International Journal of Pavement Engineering, **16**, (2015)
16 M. Słowik, M. Bartkowiak, Road engineering, **7-8**, (2016)

Protocols from periodic inspection of buildings in text mining analysis - advantages and problems of analysis

Marcin Gajzler[1]

Poznan University of Technology, Institute of Structural Engineering, Piotrowo 5 Poznań, Poland

Abstract. The article presents the problem of the analysis of post-inspection protocols of buildings in order to obtain the necessary data and information for the construction of a system supporting the technical management of residential buildings. The attention was primarily focused on the document itself - a protocol of control and automated technique of obtaining data and information from this document - text mining technique. Problems appearing in the analysis and the potential for their removal were indicated. The conclusions formulated relate to improvement of the process of conducting the review and formulating the post-control protocol in the aspect of its use in automated data collection and information transfer involving state of buildings.

1. Introduction

The subject of the article results from a broader issue that is currently being implemented - the construction of a system supporting the technical management of buildings with the use of available knowledge resources. The problem of building management has been developed for a long time as an element combining science and practice [1, 6, 12]. The technical conditions of the building in the maintenance policy and technical building management is a matter repeatedly undertaken in scientific works [4, 5, 8, 12]. The system being under construction is responsible for supporting decisions related to shaping the policy of maintenance, renovation and modernization of residential buildings. It is dedicated to entities dealing with real estate management - property managers. In the course of the conducted research, it

[1] Corresponding author: marcin.gajzler@put.poznan.pl

turned out that an important management problem is the planning of renovation policy with a limited availability of funds. Long-term planning of activities is necessary - in the perspective of several years, in order to prepare for them, and in particular to provide the necessary financial resources. It is helpful in this respect to forecast both the technical condition, the necessary expenses as well as the consequences of choosing a particular method of action in the maintenance policy. One of the knowledge resources that are used in the construction of the technical management support system are documents that occur during the operation of the building and are associated with the implementation of compulsory periodic inspections of technical condition of buildings in Poland. In the course of control of construction works, the following factors are assessed: essential elements of the construction of the facility, installations and selected technical equipment. As a result of the inspections carried out, a protocol is created describing the current technical condition and formulating recommendations in relation to the irregularities found. This protocol is an element building object book and is in the form of a text document, therefore, for the purpose of its analysis, the proposed text mining analysis is justified.

2. Basics for data and text mining analysis.

Mining analyzes compared to classical statistical methods are relatively young techniques having roots in the techniques of knowledge discovery and machine learning. Interestingly, in the course of using the mining approach, it is also possible to use classical statistics techniques. Mining techniques rely on the processing of potentially used data in order to find new interesting and often accidental dependencies. The mining approach, taking into account the results of the analysis, can be divided into targeted and non-targeted - accidental. Text mining is a younger version of the data mining date based on the analysis of text documents (data). It involves the development of a formal / quantitative representation of text data, which are subsequently analyzed using various available techniques (including statistical methods). The general text mining algorithm is presented on Figure 1.

A wider look at the methods and possibilities of conducting analyzes was presented in [11]. On the other hand in 2010[2] the application of text mining methods in the analysis of issues related to technological decision support in construction were presented. It can be stated that the suitability of the method has been demonstrated, but also some important limitations have been pointed out, in particular the sensitivity of quantitative data to the analysis and a complete loss of their context in the case of text mining analysis.

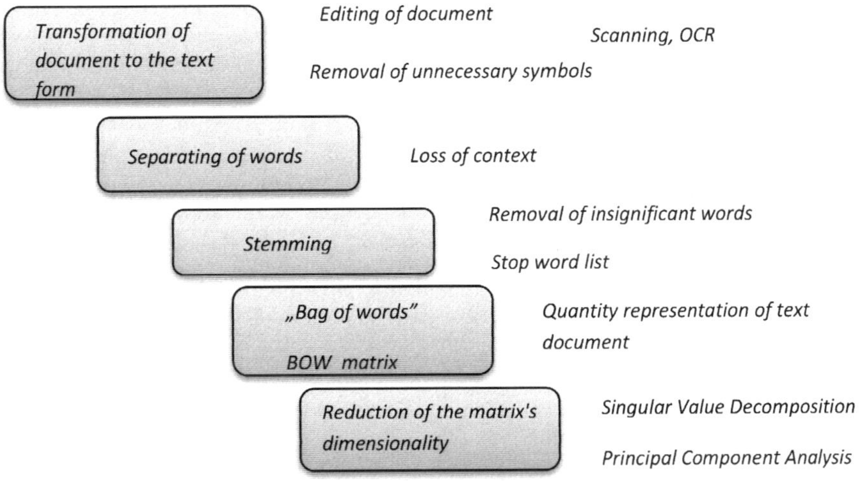

Fig. 1. The basic algorithm of the text mining process\

3. Control of building objects

The control of construction works is primarily due to the legal provisions in force in Poland and applies in particular to multi-family residential buildings. This obligation rests with the owners or established managers of the property. The inspection is entrusted to a person with appropriate qualifications to perform independent technical functions in construction. Checks are carried out annually, and expanded inspections are carried out every 5 years. The scope of control of construction facilities is described, among others in [1, 3]. Apart from the fact that the inspection obligation is imposed by the law in force in Poland, the inspection of the technical condition of the building itself is important from the point of view of conducting a proper maintenance policy of the building and guaranteeing the proper technical condition of the building. The results of the inspection allow for monitoring the technical condition of the building and in the event of the first signals of any irregularities to plan the taking of appropriate actions. At the same time, it should be noted that in fact, the necessary steps to maintain the proper technical condition are limited by the availability of funds (primarily financial), and therefore the maintenance activities are correlated with the prior acquisition of funds for necessary actions.

The post-inspection report is the result of the conducted technical inspections. It is a study containing general characteristics of the building undergoing inspection, taking into account basic surface-cubic parameters, construction and technological parameters as well as the age of the building. The basic element of the post-inspection protocol is the description of the technical condition of individual building elements starting from the foundations to the roof covering. It is not possible to precisely specify elements of the structure, because it should be borne in mind that not all elements are directly visible. In addition, individual elements related to a specific building are subject to the assessment of the technical condition.

Description of the technical condition is carried out by verbal description, where the symptoms of possible degradation occurring on the assessed element are mentioned, such as cracks, scratches, deflections, corrosion phenomena. You can also find a description of the possible causes of these degradation symptoms. In addition, a qualitative assessment of the element is made using often any dictionary, eg good, bad, satisfactory, and also recommendations for the assessed element are made. During analysis of post-inspection reports, the quantitative assessment of the technical condition by defining the degree of technical wear expressed can be also met. As a rule, in the protocols approach for estimation of the degree of technical wear, is not set, although the most common approach is based on a symptomatic visual assessment. In addition to the above-mentioned elements, the post-inspection report may contain a number of additional descriptive information, eg concerning the maintenance and repair works carried out in the last period and their impact on the current technical condition. The photo service complements the post-inspection protocol.

The analyzes of post-inspection protocols of 40 buildings from the last 10 years carried out in the course of the research indicate that these protocols are not standardized in terms of both the form and the dictionary used to describe the technical condition.

4. Text mining analysis – possibilities and barriers

The use of the text mining method in the analysis of periodic inspection reports of buildings in the context of the construction of a system supporting the technical management of residential buildings is justified and necessary at the same time. Post-inspection reports are the basic source of information on the technical condition of the building. It would be unreasonable to omit such vital information about the building. The problem that requires the use of the text mining method is that the description of the technical condition of the building in the post-inspection report is represented by a text description. In addition to it, there are also often quantitative information regarding the technical condition (eg technical wear of the facility), however, due to the lack of standardization protocols do not occur each time, and also do not allow for in-depth knowledge of factors conditioning the value presented in the protocol. The text description of the technical condition allows for a broader understanding of both the factors affecting the building, symptoms and possible directions of action.

Of course, the significance of quantitative information should not be marginalized, since these, combined with other available information, allow to learn about interesting relationships. An example of such are the dependencies of expenditures in subsequent years and the technical condition characterized by the degree of technical wear. Analyzing the mentioned dependencies of expenditures and technical wear of the building, some recurring models are noticed. In most of them, a typical relationship is already observed, namely the increase in maintenance costs causes in the perspective of one to three years the improvement of the technical condition and the reduction of the determined degree of technical wear. However, this does not apply to all objects, and there are also situations in which the occurrence of outlays does not cause a noticeable change in the technical condition of the building. This is the case in which, either the inputs take inadequate quantities or occur too late. Based on these dependencies, one can infer about proper or incorrect maintenance policy, which is an element of technical building management.

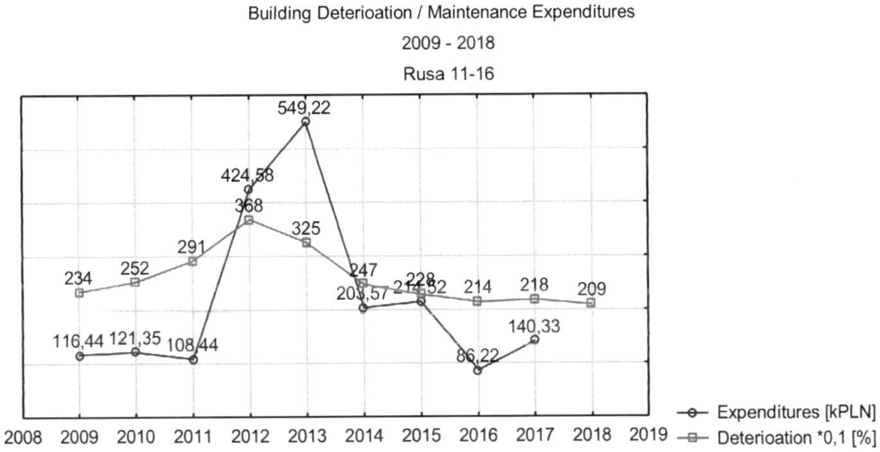

Fig. 2. Example of dependence of maintenance expenditures and the degree of technical wear for a residential building in precast concrete technology from 1979

However, focusing on the analysis of text mining, it is worth commenting on the possibilities offered by the analysis and the barriers and limitations associated with it. The author's first experiments with text mining analyzes were presented in []. Then they were used to support technological decisions. Text documents analyzed at that time constituted technical cards for various building materials. A characteristic feature of these cards was the occurrence of various quantitative data in the text description (eg strength values in MPa, application and bonding times in minutes, application temperatures in degrees of Celsius). The post-inspection reports analyzed today are much poorer in quantitative values. They occur sporadically or not at all. The dominant element of the post-control protocol is a pure text description. Even if in the technical condition description there is a need to characterize some measurable value, eg deflections, damage - scratches, cracks, due to the fact that the description of the technical condition of the building is a derivative of the inspections performed, measurable values are usually characterized by qualitative variables - fuzzy (eg permissible, over-standard deflection, deep scratches, etc.). Potentially such a qualitative description - fuzzy may seem like a weak point in technical issues, in the case of the text mining method allows for a more complete consideration of the information contained in the analyzed description. The above conclusion will be presented on the example of a text description fragment. The first fragment was taken from the post-control protocol in which the quantitative data occurred, while the second fragment from the more common plain-text description.

Fragment 1:

„Cracks in concrete approx. 3 mm between the side wall of balcony and the wall of the building. Contamination of supports by algae. Measured deflection of the 7 mm bracket. Cracks on the joints of the cover plates".

Fragment2:

„On the basis of the lack of cracks in the bearing walls, it can be assumed that the condition of the foundations is correct. Negligible scratching of selected load-bearing walls in the common part. It has been assumed that the condition of load-bearing walls (internal) and their joints are good".

Classical analysis of both fragments allows to obtain the right information, and most importantly, the correct interpretation of the phenomena described. It turns out that in the case of text mining analysis, both fragments can be included with different results in terms of information. The decisive element here is the occurrence in the first fragment of numerical values which in the course of the text mining process lose their context and informative meaning. This is due to the fact that in the course of the text mining process, individual parts of the text are subject to various processes of machining, in particular separation, reduction and steeming. Often numerical values are reduced as a symbol (not text), and if they even go through the text mining process, they are an abstract element that does not carry any information load. Separated from units and names of values are devoid of any context and, at the most, they introduce unnecessary noise. As a result of the text mining process in the basic approach, a frequency matrix is obtained, which counts the frequency of the extracted and processed words. The frequency matrix is also difficult to interpret in terms of the information originally conveyed by the text, but the awareness of individual components of the text mining process allows categorically to state that numerical values lose their context and information load.

The frequency matrix is also the first basic quantitative representation of the text. Taking into account the length of the text description, it can be expected that the frequency matrix will be characterized as oversized. To reduce it, the decomposition of singular values is used. The representation thus obtained is already a complete abstract representation of the text. However, awareness should be paid that despite the abstract form, the information load is contained in this form. Taking into account the building management system for building management, the text mining analysis allowed obtaining the BOW frequency matrix with the dimension 400x273 (the number of rows results from the number of buildings in the database and the adopted 10-year analysis period, while the number of columns concerns words included in the analysis resulting from the text mining text documents). After applying the decomposition, a matrix of 400x5 can be obtained, where each of the 400 rows is responsible for representing the technical condition of a specific building in a given year according to the established nomenclature:

019: [2,3413 1,4193 1,5616 2,4809 3,1036]
018: [1,9803 1,4232 1,9425 2,0377 2,1807]
....
....
040: [2,5501 1,7052 3,1428 2,3882 1,3209]

where the first two digits of the row number correspond to the number of the building, and the third number is denoted to the time in years back from the year of analysis, eg the number 9 meant 2009, while the correct matrix was peculiar values representing the technical condition for a given building and a given year of its operation.

Taking into account the technical building management system being built, the technical condition is one of 11 variables (building year, building technology, usable floor space, number of floors, technical condition, maintenance costs, ownership structure, number of owners, ownership share of the city, conservation supervision) included in the analysis. Due to the use of quantitative and intelligent methods available in the group of data mining methods, it was necessary to bring a description representing the technical condition to an acceptable form in the presented analyzes. The set of cases characterized by the 11 variables creates a base used for CBR [7, 13] for new cases. The case selection mechanism from the database includes several variables, including similarity of technical condition. For this purpose, agglomeration methods have been used, in particular the Ward method [3]. However, the use of this method is justified only after prior preselection of cases from the database and is the last element of selection. Empirical methods have shown that the building selection is a selective variable. It is so important that, as shown by a cursory analysis of the case database, it is strongly correlated with the year of construction of the building. This is confirmed by the development of housing construction in Poland, which can be divided into brightly visible stages: traditional construction, prefabricated construction and a return to traditional and mixed technologies.

5. Conclusions

Conducted in connection with the construction of a system supporting the technical management of residential buildings the analysis of test documents - post-inspection protocols pointed to numerous problems and barriers to the application of the analysis. The first group are technical problems regarding the analysis itself. The software on which the Statistica Data Miner is run does not have any dictionaries implemented to allow the use of the steaming process for documents in Polish. This makes it necessary to conduct this part of the analysis in a manual way. There are of course other software solutions (PS Clementine PRO and Applica MPJ module), but these are not so extensive in the field of analysis methods. Another problem in text-based analysis is the occurrence of quantitative data in text documents that lose context and information in the course of the process. A very important problem regarding the text documents themselves is the lack of their standardization in terms of the information contained therein as well as the form itself and the dictionary used. The proposal solving most of the problems mentioned above is the construction of a universal form allowing for the standardization of both the description of the technical condition of the building and the available description dictionary. It seems even legitimate to build a mobile application available for devices such as smartphone and tablet, which using a fixed form and an accessible dictionary will allow you to generate a post-control form and simultaneously obtain a text document - standardized in electronic form, ready to conduct analyzes. Such approach would also allow dynamic and significant development of the available case database of the management support system.

References

1. Bucoń R., M. decyzyjny wyboru wariantów remontu lub przebudowy budynków mieszkalnych. Politechnika Lubelska, 2017
2. Gajzler M., TEDE, **16** (2010), *219-232*
3. Gajzler M., *Proc.Eng,*. **208** (2017), *20-26*
4. Halim, T., Muthusamy, K., Chia, S.Y., Lam, S.W.: Facilities, **29** (2011) , *114-132*
5. Horner, R.M., El-Haram, M.A., Munns, A., Int. J. of Quality in Maint., Eng. **3** (1997) , *273-280*
6. Kapliński O.,. Eng. Struct. Techn., **10**, (2018), *27-33*
7. Kończak A., Pasławski J., *Proc. Eng.* **57** (2013), *583-588*
8. Moghaddam, K.S., Usher, J.S., J. of Quality in Maint. Eng., **16** (2010) 3, *271-287.*
9. Nowotarski P., Paslawski J., Lean and agile management synergy in construction of high-rise office building. Archives of Civil Engineering, ***62*(4),** (2016), 133-148,
10. Plebankiewicz E., Zima K., Wieczorek D., Archiv. of Civ. Eng., **62** (2016), *149-166*
11. Shu-Hsien Liao, Pei-Hui Chu, Pei-Yuan Hsiao, Exp. Sys. with Applic., **39** (2012) *11303-11311*
12. Tsang, A. H. C.: J. of Quality in Maint. Eng., **1** (1995) *3-17*
13. Zima K.. *Proc. Eng.*, **122**(2015), *57-64*

Flexible procedure for concreting in hot and dry climate based on Phase Change Material

Mahmoud HSINO[1,*] and *Jerzy PASLAWSKI*[2]

[1] Stanislaw Staszic University of Applied Sciences in Pila, Politechnic Institute, 64-920 Pila, 10 Podchorazych Str., Poland
[2]Poznan University of Technology, Department of Construction and Environmental Engineering, 60-965 Poznan, 5 Piotrowo Str., Poland

Abstract. The article presents the possibilities of using a flexible approach in designing production processes in construction on an example of concreting in a hot and dry climate. The basic drawback of typical quality management system procedures is their rigidity. It is based on the assumption that building processes are carried out in favorable conditions. Meanwhile, these conditions often change to a large extent during the process, and the accuracy of weather forecasts is limited. The aim of the article is to demonstrate the advantages of flexibility procedures for concreting on the example of concrete modification with phase-change materials. The conducted research has shown a reduction in the risk of structure damage during maturing of young concrete due to the use of hybrid phase-change materials in the conditions of central Syria climate. Phase-change materials provide lowering of the thermal peak in the initial maturation period and unification of the temperature in the cross-section of the element (limitation of the thermal gradient). At the same time, PCM's slightly reduce the compressive / tensile strength of the concrete. As a result, this reduces the risk of cracking of the concrete element, which is the basis for ensuring greater durability of the element and a longer life cycle of the building.

1 Introduction

Building Construction due to its specificity is characterized by a significant difference in relation to other sectors of economy. Important factors related to the specificity of construction include:
- Immovability of the product
- Unstable production conditions
- Relatively high unit cost of the product

* Corresponding author: m-hsino@o2.pl

- Problems with maintaining permanent staff
- Seasonality in construction output
- Sensitivity of production to changing environment

The last two features are related to the dependence on the impact of the season. In the case of seasonality, the dependence on budget restrictions is often emphasized (usually at the beginning of the year, while at the end of the year the necessity of hastily money spending is taken into account). In turn, the second feature is related to the sensitivity of weather factors, which can lead to the extreme necessity of demolition of the concrete element and re-work. Naturally, various reasons should be taken into account when analyzing the causes of problems in the implementation of construction processes. They can primarily be classified into two categories: related to construction and resulting from the impact of the broadly understood building environment (environmental, socio-political or legal factors). The factors directly related to construction include: rising costs of salary, materials and supplies, delays or inflation, lack of qualified employees, errors in project documentation, lack of due supervision, untimely deliveries and implementation of processes in uncontrolled conditions (significant impact of weather and other external factors). The proper implementation of health and safety regulations is also associated with the correct course of construction works. The involved construction manager, thanks to proper communication between all participants of the construction process, prevents conflicts and achieves harmonious cooperation. Achieving success requires efficient risk management, taking into account many potential sources. From this perspective, the manager's support in weather risk management in concreting can be an important element guaranteeing compliance with the requirements regarding costs, time of completion and quality.

The article focuses on procedures related to the concreting process under hot and dry climate conditions, which means weather risk management in countries where high temperature, low humidity and high wind speed pose a threat to the correct construction of the concrete structure.

Weather factors have an impact on the construction industry due to the implementation of processes in a large part on the construction site. Weather risks occur outside of construction also in other sectors of the economy, such as energy (electricity demand dependent on weather) or agriculture (risk of reduced yields as a result of drought or frost).

In construction, quality management systems are often based on too rigid procedures that do not take the variability of the environment into account. As a rule, it is assumed that the processes are carried out under favorable conditions, with some restrictions on their use. For example, 13°C [1] is considered the optimum temperature for concreting and maturing. On the other hand, the borderline upper limit of ambient temperature for concreting is often assumed at 29-32°C. Formularizing such restrictions seems to be debatable, because the temperature of concrete is of key importance. Situations when it becomes necessary to repair a damaged element of critical infrastructure can be easily imagined (runway at the airport, bridge, drainage ditch, road, etc.). Quality management system procedures should also take into account unusual conditions when concreting is allowed under special conditions. An example is the use of phase-change materials that allow to exceed typical technological barriers due to unique properties [2, 3]. Another example of innovative solution for concreting under hot climatic conditions is the use of superabsorbing polymers [4] or manioc based retardants [5].

The aim of the article is to present the beneficial effects of the use of phase-change materials in concreting in hot and dry climates, when the ambient temperature reaches 45°C, and the daily temperature amplitude is 30°C.
The graph below (Fig. 1) of the maximum and minimum monthly air temperature in Damascus indicates the existence of significant differences between them (e.g. 35°C in July).

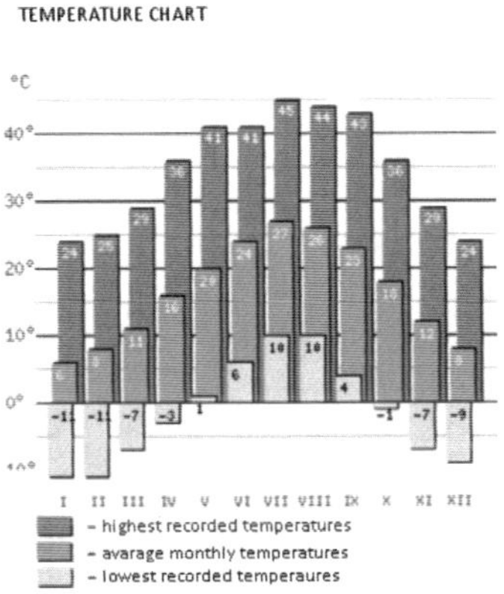

Fig.1. Maximum and minimum monthly temperatures in Damascus [6]

After this introduction, the basic principles of concreting in these unfavorable conditions were presented, and then the basic possibilities of using the technological flexibility were indicated. Later, the research procedure is described in a specially constructed climatic chamber, which enabled the preparation and testing of three concrete mix recipes. The research results and conclusions are subsequently presented.

2 Problem definition

A view is often found that concreting at elevated ambient temperature in the initial period has a positive effect on early concrete strength and does not cause adverse effects at a later time. However, as shown in the detailed initial studies, fast cement hydration contributes to a weaker concrete (more porous) microstructure. This hypothesis is confirmed not only in strength development tests (the initial high dynamics of compressive strength build-up is significantly slowed down [7]), but above all in a negative impact on the durability of concrete [8, 9]. The cause of these unfavorable phenomena is the fact that at a high initial rate of hydration there is too little time for ion diffusion outside the layer of hydration products directly surrounding the cement grains. This limits the homogeneous loss of hydration products in the space between the grains, which occurs at a lower temperature [10]. The basic problem in these maturation conditions is the risk of exceeding the tensile strength of the concrete by thermal stresses in freshly-cured concrete. Phase-change materials enable the reduction of the thermal peak in the initial period of maturation occurring due to the exothermic reaction dynamized by the high temperature of the environment and provide a more even temperature distribution in the cross-section of the ripening element (reduction of the thermal gradient). The possibilities of using phase change materials for this purpose were indicated by Bentz and Trupin [11]. The simulation tests carried out by Arora et al. confirmed the validity of these assumptions [12] and laboratory tests. Savija [13] showed the benefits of using phase change materials also during operation (lower maintenance costs of infrastructure facilities – e.g. bridges).

3 Hot weather concreting procedure

Principles of laying concrete in high temperature can be included in the following 10 points [1, 14]:
1. Ensuring adequate possibilities of collection, laying and care of concrete at the construction site
2. Using larger aggregate fractions
3. Not laying concrete in the afternoon or directly in the afternoon
4. Planning to deliver a concrete mix taking into account its reduced temperature (adding ice, spraying the aggregate with water, etc.)
5. Agreeing with the designer of the construction of expansion joints (more often than at a lower temperature)
6. Using canopies and wind shields to reduce potential problems resulting from direct sunlight or increase evaporation of concrete water due to wind
7. Applying the appropriate implementation schedule - for example, concreting the floor after making the walls and roof
8. Application of measures to limit evaporation
9. Using ice as part of the water dispensed into the concrete mix or liquid nitrogen to remove the concrete components
10. Moistening the substrate before concreting.

Proper curing of the fresh concrete plays a key role.

In Fig. 2 different concrete curing methods are listed, also considering the use of phase-change material as an internal thermal effect.

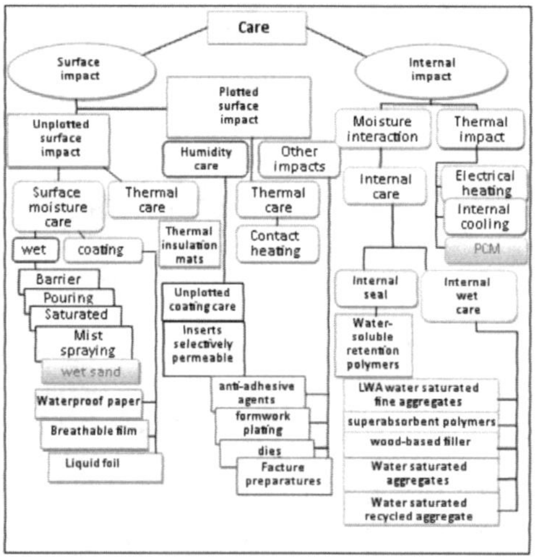

Fig. 2. Different curing procedures including phase change material (PCM) – based on [15].

From the point of view of this study, the methods of cooling the concrete components, the use of retarding additives, the use of cement with low dynamics of hydration heat release, and in the scope of care methods:
- In the case of an unclosed surface - the use of concrete preservatives in the form of a protective film on the surface of freshly poured concrete (the main advantage is the possibility of applying on large surfaces practically immediately after laying the concrete mix) or applying a layer of damp sand on the concrete surface providing periodic wetting with water (if the surface is horizontal)
- In the case of a decorated surface - the use of external shields to prevent evaporation of water from the concrete surface
- The use of phase change materials as an internal curing option (possible in both cases for the care of the unclosed and closed surfaces). The aim of the research was to compare the risk of damage to the concrete structure in the initial maturation phase based on the determination of the cracking risk coefficient:

$$Cr = \Delta T / fc \quad (1)$$

where: Cr - Cracking risk coefficient ΔT - temperature gradient [°C] fc - compressive strength of concrete [MPa]. Naturally more precise risk of concrete cracking can be determined by following thermal stress and tensile strength. The present research was aimed at confirming the use of hybrid phase change materials and optional modification with a retarder.

4 Laboratory tests using phase change materials .

In order to carry out the experiments, a climatic chamber was built, which allows testing different concrete mix in hot climates. The aim of the research is to determine the effect of PCM - (Phase Change Materials) on the reduction of thermal gradients inside a concrete slab in the initial maturing phase. The chart below presents the daily cycle of temperature and humidity (from 20:00 to 20:00 the next day), prevailing in hot and dry climate (Hama - Syria).

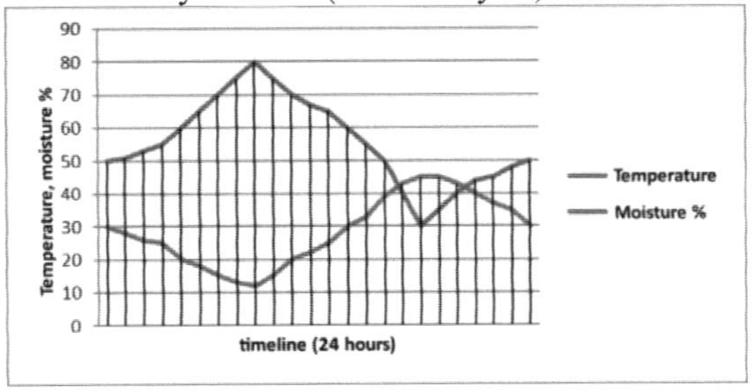

Fig. 3. Diurnal course of temperature and humidity cycle in the summer, in Hama:(source: own)

As part of the experiment, three plates (samples) of 50x50x20 cm were made. In this experiment, i.e. a hybrid solution was used, namely two different phase change materials (PCM) and a retarder were used.
- The first plate is composed of concrete mix with two PCM materials and a retarder, in short - RET

- The second plate is composed of concrete without any additions (witness), in short - NOR
- Third plate is composed of concrete with the addition of two PCM materials, in short 2 PC

The first material (PCM) has following characteristics:
- Phase transition temperature of 23 ÷ 26 °C
- Thermal capacity 142 kJ / kg
- Addition rate of 2.4% in relation to cement dose

The second one is characterized by:
- Phase transition temperature from 29°C to 36°C
- Heat storage capacity 160 [kJ / kg]
- Addition rate of 2.4% in relation to cement dose

Table 1 presents the recipe of the plate components – RET, concrete class C 30/37, consistency S4.

Table 1. The composition of the concrete slab mixture – RET

Material name	**Supply source**	**The amount of kg per 1 m³ of compacted concrete**	**Density kg/m³**
CEM II B-V 42,5N	Warta	21,56	3,1
Plain sand 0/2	Dabrowa	54	2,65
Gravel 2/8	Dabrowa	28,125	2,65
Gravel 8/16	Dabrowa	28,125	2,65
Retarder FL	SIKA	0,151	
Plasticizer Plastiment BV IM	SIKA	0,129	
Water	Water supply	12,27	
PCM -1		0,518	
PCM- 2		0,518	

Table 2 presents the recipe of the plate components – NOR, concrete class C 30/37, consistency S4.

Table 2. The composition of the concrete slab mixture - NOR

Material name	**Supply source**	**The amount of kg per 1 m³ of compacted concrete**	**Density kg/m³**
CEM II B-V 42,5N	Warta	21,56	3,1
Plain sand 0/2	Dabrowa	56,25	2,65
Gravel 2/8	Dabrowa	28,125	2,65
Gravel 8/16	Dabrowa	28,125	2,65
Retarder FL	SIKA		
Plasticizer Plastiment BV IM	SIKA	0,129	
Water	Water supply	12,94	

Table 3 presents the recipe of the plate components - 2 PC, concrete class C30/37, consistency S4.

Table 3. The composition of the concrete slab mixture – 2 PC

Material name	Supply source	The ammount of kg per 1 m³ of compacted concrete	Densitykg/m³
CEM II B-V 42,5N	Warta	21,56	3,1
Plain sand 0/2	Dabrowa	54	2,65
Gravel 2/8	Dabrowa	28,125	2,65
Gravel 8/16	Dabrowa	28,125	2,65
Retarder FL	SIKA		
Plasticizer Plastiment BV IM	SIKA	0,129	
Water	Water supply	12,42	
PCM -1		0,518	
PCM- 2		0,518	

5 Results

The graphs of temperature waveforms in individual plates are presented below.

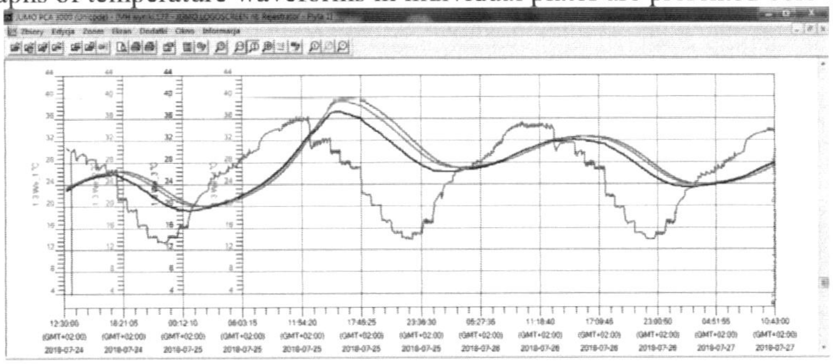

Fig.4 . The temperature course in the RET plate (source: own)

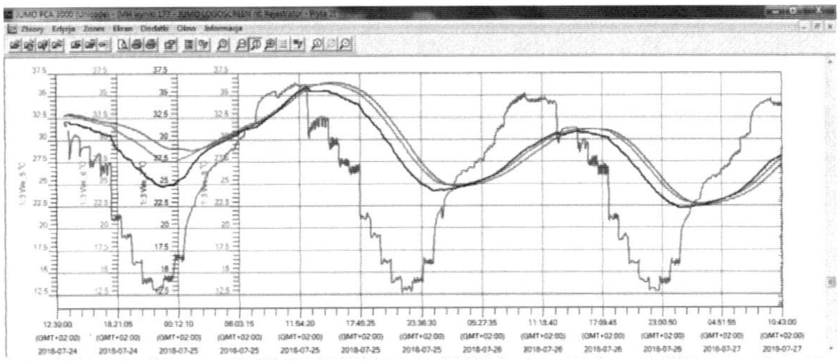

Fig.5. The temperature course in the NOR plate (source: own)

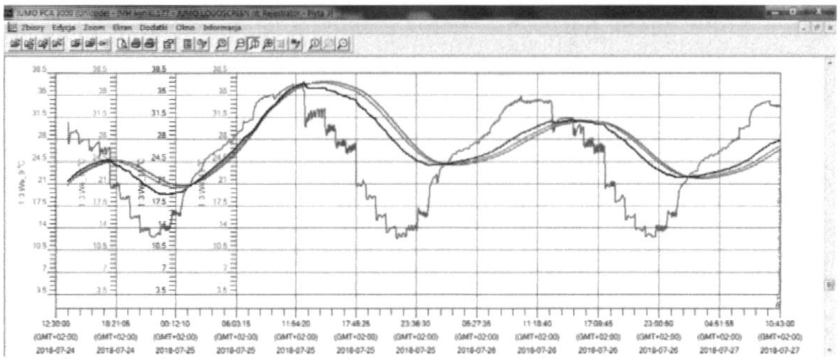

Fig.6. The temperature course in the NOR plate (source: own)

At the same time, from the same concrete mix from which the plate was made, 3 cubes for each mix of 15x15x15 cm were made. These cubes were made to check the course of compressive strength over time and to determine the risk of cracking of the concrete structure in individual plates in the initial period of maturing.
Based on the obtained results, it can be noticed that the risk of cracking the concrete structure in a board without additives (NOR) is almost one and a half times higher than in panels where phase-change materials (2PC) were added.
The following results are obtained on the basis of the first (1) calculation.

$$C_r^{RET} = \frac{2,7}{0,2} = 13,5 \tag{2}$$

$$C_r^{2PC} = \frac{2,5}{3,25} = 0,77 \tag{3}$$

$$C_r^{NOR} = \frac{4,3}{4} = 1,08 \tag{4}$$

The obtained results confirm a slow increase in compressive strength in the slabs with the addition of PCM, which means that the structure of the mature concrete will not be defected, as mentioned at the beginning of the article.
All slabs were placed on the same height to eliminate difference in temperature between the individual boards.
Photo number 7. Presents the view of samples in the climatic chamber

Fig 7. View of samples in the chamber

6 Conclusions

This research and theoretical considerations allow to draw the following conclusions:

1. Phase-change materials used for modification of cement concrete allow to overcome typical technological barriers in the case of concreting in unfavorable conditions (hot and dry climate)
2. It is justified to apply technological flexibility in the form of several curing options depending on the conditions of concreting and maturing, technological capabilities of the plant and the type of concrete element
3. The use of PCM as a cement concrete modifier allows limiting the size of the thermal peak and unification of the temperature in the section of the concrete element, which leads to a reduction in the risk of cracking (which may disqualify concrete element)
4. The addition of PCM to concrete also limits the amount of temperature changes during the operation of the concrete element, which may lead to increased durability and extension of the life cycle of the concrete element
5. The conducted research shows that the use of hybrid solutions based on two phase-change materials reduces the risk of cracking of the concrete structure during its maturation process.

Further research will aim at building a simulation model with experimental verification.

Acknowledgements

The authors wish to thank for finance support from the Institute of Building Structures at Poznan University of Technology allowing to perform laboratory tests as part of the statutory activities.

Reference

1. A. M. Neville, *Wlasciwosci betonu* (Krakow, 2012)
2. A. Y. Benjamin, F. Gabriel, S. Zhenyu, M. T. Alexander, W. Zhenhua, N. Narayanan, S. Gaurav, L. Pilon, Const. and Build. Mat. **147**, 466–477 (2017)
3. H. Ahmed, A. A. Khaled, A. Hamza, R. Yasir, A. Shaimaa, Eff. of PCM Int. Conc. Bl. H. G. Prev. H. Clim. **8**, 1009 (2016)
4. M. Hasholt et al. Const. Build. Mater. **31**, 226-230 (2012)
5. G. R. Otoko. Eur. Intern. J. Sci. Tech. **3**, 2 (2014)
6. www.transazja.pl
7. R.G. Burg, *The Influence of Casting and Curing Temperaure on the Properties of Fresh and Hardened Concrete*, (Portland Cement Association 1996)
8. O. Idowu, L. Black, Mag. Conc. Res. **70**, 633-647 (2018)
9. N. Gowripalan, J. Cabrera, A. Cusens, P. Wainwright, Conc. Inern. **12**, 47-54 (1990)
10. M. Hsino, J. Paslawski. Izol. **2**, 46-49 (2014)

11. D. P. Bentz, R. Trupin, Cem. Conc. Comp. **29**, 527-532 (2007)
12. A. Arora, G. Sant, N. Neithalath, Cem. Conc. Comp, **81**, 11- 24 (2017)
13. B. Savija, Mat. **11**, 654 (2018)
14. ACI 305R-10, *Guide to Hot Weather Concreting* (American Concrete Institute, 2010)
15. P. Woyciechowski, W. Jackiewicz- Rek, *Materialy Budowlane*, **5**, 44-48 (2012)

Scientific activities in Construction Technology Management filed of Construction Innovations Academic Association

Marcela Jasińska[1]*,

[1]Poznan University of Technology, pl. M. Skłodowskiej-Curie 5, 60-965 Poznan, Poland

Poznan University of Technology is a university with a technical profile, with 10 faculties, including the Faculty of Civil and Environmental Engineering. The university is also open to developing interests and passions of students, through the existence of numerous scientific circles. Officially, there are 74 academic and 15 student organizations. Among them is the Scientific Circle Constructon Innovations, whose history dates back to 2010, although it has been officially started in 2011. It associates students not only of Polish origin, but also from distant countries, such as: Democratic Republic of Congo, Iraq or India. Its main goal is to promote the Poznan University of Technology, Faculty of Civil Engineering and Environmental Engineering and the specialties of Construction Technology Management in the student environment and in the nationwide construction industry. From year to year there is an increase in its members The article confirms that the activity of scientific clubs has a very positive impact on the development of students' interests. Confirmation of this thesis, on the example of ConInno are research projects from year to year, with a variety of interesting topics

1 Introduction

Scientific activities play very important role in the academic associations because they strengthen the need for search for new ideas, methods and problem solving for subjects that are related to the knowledge provided in the course of studies at the universities. Not only the activities itself but also participation in a student group for young people is a good way to ignore position and power hierarchy, which are replaced by collegiality and dedication to achieve both academic and professional success and enable creative thinking[1] This is especially important since for many students, education does not simply mean obtaining a university degree and, consequently, a good job. There is also a need to attain personal satisfaction and a sense of responsibility for the community in which one lives [2] What is more, formation of the personal orientation and readiness for professional work at the future engineers should be conducted in conditions of innovative educational activities [3],

* Corresponding author: marcela.jasinska@student.put.poznan.pl

which confirms the necessity to take scientific actions by academic associations at the universities.

2 Scientific associations at Poznan University of Technology

Poznan University of Technology is a state-owned university with a technical profile, located in Poznań. It has 10 faculties, 27 fields of study and over 100 specialties. Among them, there is also the Faculty of Civil and Environmental Engineering, with majors such as: Construction, Environmental Engineering and the newly created Sustainable Building Engineering- Sustainable Construction. Over 20,000 students study first and second cycle, doctoral studies and postgraduate studies, which education is provided by over 1200 academic teachers. The Poznan University of Technology as the first of Polish universities was admitted to the group of CESAER members (Conference of European Schools for Advanced Engineering Education and Research) - a European organization associating the best technical schools. He is also a member of SEFI (Societe Euro peenne pour la Formation des Ingenieurs), EUA (European University Association), ADUEM (Alliance of Universities for Democracy) and IAU (International Association of Universities).[4]

Students choose studies at the Poznan University of Technology due to the high level of teaching, perfectly prepared staff, as well as the opportunity to fully realize their interests in actively operating scientific circles. These organizations are aimed at bringing together students whose main goal is self-education and implementation within the chosen subject.

At Poznan University of Technology there are officially 74 scientific circles and 15 student organizations. Among them there are 7 scientific circles belonging to the Faculty of Civil and Environmental Engineering, also the Scientific Circle of the Poznan University of Technology: Construction Innovations (Con-Inno)[5,6]

2.1 Con-Inno academic association

The Con-Inno Scientific Circle aims to promote the Poznan University of Technology, Faculty of Civil Engineering and Environmental Engineering and the specialties of Construction Technology Management in the student environment and in the nationwide construction industry. During circle members' meetings, among others, the following issues: flexibility in construction, sustainable and passive construction, management of the construction process, technology in construction, team management, software in construction. The history of ConInno dates back to 2010, when it was the idea to create it. The circle was officially registered on December 22, 2011 with the help of dr hab. Eng. Jerzy Pasławski.

An important distinction for the ConInno Scientific Circle is the fact that it took 10th place in financing competition at the Poznan University of Technology (among the 51 Scientific Teams), thanks to which it received high budget for 2018. The circle is also ranked 1st among all the Scientific Teams of the Faculty of Civil and Environmental Engineering based on the same rank.[7]

3 Recent research projects in Con-Inno

Every year ConInno conducts many important research projects. These scientific undertakings are characterized by a large involvement of students. The submitted theses always find scientific proof. Confirmation of general assumptions of ConInno's activity are conducted in 2017 - 4 research projects [9], the course of which is described below.

Fig. 1. Logo of Construction Innovations Academic Association [8]

3.1 Accuracy of BLE systems for the H&S improvement in construction

It is a research project carried out by three members of the circle. Its effect is to complete the Master's thesis with two participants and research for the doctoral thesis. On the basis of the research, a publication [10] was also created, which was presented at the 2nd International Workshop on flexibility in sustainable construction conference in Poznań, on 24-25.04.2017

The subject of the project is primarily a problem related to the exposure of health and safety of workers in the construction industry. There are many solutions on the market that are struggling with this problem, one of them is BLE localization technology. The authors of the study decided to investigate Bluetooth devices and perform a range test on devices received from a European manufacturer. The navigation signals are based on Bluetooth Low Energy (BLE) technology, which is a medium-quality data transmission standard for low-cost devices with long-life batteries. The results turned out to be promising, and the effect of the work was used in the course of further research work of other members of the circle

3.2 Quality Improvement of Ground Works Process with the Use of Chosen Lean Management Tools–Case Study

Another research project created by a two-person team. The result of the project was the completion of a master's thesis through one member and a dissertation. On the basis of this project, a publication [11] was also created, presented at the conference World Multidisciplinary Earth Science Symposium, in Prague, on 11-15 September 2017

The main theme of the project is the introduction of selected Lean Management tools to improve the quality of the earthworks process. These tests were carried out on the basis of an analysis carried out at the construction site of a vehicle inspection station, which is located in Poznan. Process assessments were made on a different basis, assuming that all three main groups of employees were directly involved in the process. They were: manual workers, construction manager and construction engineer. What's more, the studies took into account and compared 3 points of view on problems that could have occurred during these types of work, with the analysis of details and the reasons for their creation. The authors also checked the change of approach of workers directly involved in the mentioned processes concerning the introduction of the Lean Management methodology. It illustrates the problem of people's scepticism towards new ideas who are used to using traditional methods while working.

The introduction of Lean Management tools is especially important because earthworks are one of the first processes during the construction process. Various types and solutions for foundations were developed depending on the ground conditions. They are the type of soil or groundwater level. Well and solidly made foundations are the basis of the building, and well-made guarantee a long life of the structure and guarantee smaller, future costs associated with possible repairs.

3.3 Formal and Legal Aspects of Buying and Commissioning Flats

A research project created by a three-person team. Its effect was the use of data for a doctoral dissertation. On the basis of the project, a publication [12] was also created, which was presented at the 2nd World Multidisciplinary Civil Engineering Conference - Symposium Urban Planning, in Prague, on June 12-16, 2017.

The subject of the research are formal and legal issues and analyses accompanying the procedure of purchase and collection of apartments in Poland. This is a very current topic and concerns a wide group of buyers. Every year in Poland I sell and take away many flats. However, not all buyers are aware that its purchase requires not only knowledge from the construction side, but also legal aspects. The project speaks primarily about the rules related to the purchase of a flat, the process of acceptance of works, removal of detected defects, failure-free receipt or enforcement of a guarantee. The author also presents a logical connection between the three sides or their indirect and direct dependencies. The parties are the buyer, developer and general contractor. The survey also presents all kinds of problems resulting from delays in completion of work due to the developer's fault and talks about possible legal paths to claim their rights.

3.4 Review of the Air-Coupled Impact-Echo Method for NonDestructive Testing

A research project prepared by a three-person team. Its effect is also the implementation of research for the doctoral thesis and preparation of the publication [13]. It was presented at the 2nd World Multidisciplinary Civil Engineering conference.

The main topic is the concept of the Air-Coupled Impact-Echo (ACIE) method. It is one of the non-destructive testing (NTD) techniques that is used in construction. The main advantage of Impact Echo (IE) is that you only need access to the structure from one side of it. This facilitates testing in road facilities and places where access is difficult

The aim of the research project is to provide knowledge about the Air-Coupled Impact-Echo method based on a publication available at the Thomson Reuters Web of Science Core Collection database (WOS). From the analysis, it was concluded that ACIE methods can be widely used to perform non-destructive testing on concrete structures. This method is also faster than the IE method thanks to sensors coupled with air.

4 Conclusions

The data presented above on the activities of the academic association - Construction Innovations last year at the Poznan University of Technology prompts the following conclusions:

1. Interesting subject of works carried out in cooperation with the scientific circle arouses from year to year growing interest among students of the Poznan University of Technology.

2. The Con-Inno in 2017 carried out 4 large research projects whose effects were reflected in the form of scientific publications and promotional works of students implementing them.
3. Con-Inno was appreciated for its scientific activity, taking the 10th place among all academic associations at the Poznan University of Technology and the 1st place among the active student groups at the Faculty of Civil and Environmental Engineering.

The above article confirms that the scientific circles have a very positive impact on the development of Poznan University of Technology students. They have the opportunity to find answers to bothering questions and to solve problems in their field of interest.

References

1. V. Yepes, J.V. Marti, T. Garcia-Segura, Desarrollo y evaluación de la competencia transversal "pensamiento crítico" en el grado de ingeniería civil. (In-RED, 2016)
2. J. C. Borges, T. C. Ferreira, M. S. Borges de Oliveira, N. Macini, A.C. F.a Caldana, a, Student organizations and Communities of Practice: Actions for the 2030 Agenda for Sustainable Development, The International Journal of Management Education 15(2 B), 153-161 (2017,)
3. S. Maksimova, Z. Kutrunova, L. Maksimov, A. Voronov, Experience in the use of modern educational technologies in teaching professional disciplines of training direction "Civil Engineering", MATEC Web of Conferences, 106, p. 09020, EDP Sciences (2017)
4. www. put.poznan.pl (accessed: 20.07.2018)
5. P. Nowotarski, J. Pasławski,, Ł. Pluciński, Scientific association knowledge improvement activities in Construction Technology Management field, Procedia Engineering 208, 106-113 (2017)
6. J. Pasławski, P. Nowotarski, R. Milwicz, S. Dubas, Improving Education of Managers in Construction Sector with the Use of EU Funds, Procedia Engineering, 161, 1043-1048 (2016)
7. Rank of Academic association ar PUT(2018)
8. http://ztiob.put.poznan.pl/d2/node/41 (acessed: 20.07.2018)
9. Ł. Pluciński, Annual rep. on the act. of Con-Inno 2017 (15.12.2018)
10. P. Nowotarski, J. Pasławski, D. Mielcarek, Accuracy of BLE systems in the H&S improvement aspects in construction, Procedia Engineering, 208, 98-105 (2017)
11. P. Nowotarski, J. Pasławski, B. Wysocki, Quality Improvement of Ground Works Process with the Use of Chosen Lean Management Tools–Case Study, IOP Conference Series: Earth and Environmental Science. IOP Publishing,95(2),022032, (2017)
12. S. Dubas, P. Nowotarski, R. Milwicz, Formal and Legal Aspects of Buying and Commissioning Flats, IOP Conference Series: Materials Science and Engineering, IOP Publishing, 245(3), 032089 (2017)
13. P. Nowotarski, S. Dubas, R.Milwicz, Review of the Air-Coupled Impact-Echo Method for Non-Destructive Testing, IOP Conference Series: Materials Science and Engineering. IOP Publishing, 245(3), 032098 (2017)

Simulation as a decision tool in construction process management in turbulent environment

Kinga Katafoni[1,*], and *Jerzy* Paslawski[2]

[1]CTM Master Graduate Student, Poznan University of Technology
[2]Construction Engineering and Management, Poznan University of Technology

Abstract. This paper is dedicated to simulation idea of construction process for decision-making aid. An aim is to improve technological decisions for concreting operation at low temperature. A risk of destruction of concrete structure is a critical problem in this situation. A sudden drop in temperature may result in disqualification of the concrete element or the entire structure. The proposed solution is based on a simulation of the process of maturing of the concrete element based on experiments in the laboratory (e.g. heat of hydration and dynamics of growth of compressive strength depending on the temperature). The case study proved that the traditional approach to the management of concreting technology (based on one 48-hour forecast) is extremely risky when operating in a turbulent and changing environment. The proposed flexible approach is based on systematically updated simulation and is more effective thanks to the possibility of activating active flexible procedures to prevent freezing of concrete in case of pessimistic scenario.

1 Introduction

Construction is a field of economic activity that is significantly different from the typical production activity carried out under factory conditions. The basic difference results from the implementation of construction processes under uncontrolled conditions. The influence of weather conditions is a manifestation of the impact of the dynamically changing environment. Variable climatic conditions in time and space generate different deadlines (and related costs) for the implementation of various construction processes and the related risk of failure to meet the assumed deadline for implementation, budget overrun and failure to meet quality requirements [1, 2]. It is necessary to emphasize the often-encountered pressure to shorten the implementation cycle, which imposes the necessity to continue the works An example of a process exposed to a significant weather risk is concreting under conditions of cold temperature, when despite the start of the process under favorable conditions there is a danger of sudden deterioration of maturation conditions (frost attack) and damage to the concrete structure at the initial stage of curing. Under these conditions, decisions regarding the adopted technology are usually made in situation of information deficit and awareness of the necessity to implement the process in an

* Corresponding author: katafoni.kinga@gmail.com

environment characterized by a large range of variability, high uncertainty and high dynamics of change.

The use of a flexible approach enables sequential decision making based on gradually updated data and simulation models for launching appropriate procedures in the event of a pessimistic scenario. It gives the opportunity to continue the process implementation while ensuring the achievement of the assumed results (quality requirements) despite unfavorable weather conditions.

The purpose of this article is to show the important role that simulation plays as a tool for managing construction processes. It is important to emphasize the importance of synergy between monitoring (information on the course of the process in progress and its environment), a gradually updated forecast and simulation. This gives the basis for decision-making at individual stages of the process.

After this introduction, a problem has been defined which involves a high risk of damaging the structure of the concrete maturing at low temperature, when a sudden drop in ambient temperature can lead to irreversible changes in the concrete structure (e.g. a decrease in compressive strength by 80 %.). This often means that you need to demolish an item and re-execute it. Based on the authors' own experience, this may lead to an increase in costs of up to 300-400%. Then, a solution based on a flexible approach was indicated.

The third part indicates the important role of simulation in the decision making process regarding the application of the active flexibility option (adaptability). The example describes the application of the proposed method in the case of concreting a gas station sealing plate at reduced temperature. The article was concluded with conclusions indicating the prospects of further research.

2 Problem definition

Climate condition in Poland are variable and changes in temperature have huge impact on processes on the building site. The most sensitive process to low temperature is concreting. Wrong estimation of weather condition generate a huge amount of waste and costs. Moreover it effects delays in schedule due to fact that stripping time for the concrete form work has significant influence on schedule. If the process of form work removal is performed too soon there is a possibility to destroy the construction. Furthermore, it creates a hazardous condition for workers. The Table 1 shows the dependence of variable weather condition on construction works.

Table 1. Dependence of variable weather condition on construction activities [1,2].

Construction operation/ Weather condition	Earthworks	Concreting	Assembly construction
Temperature $\leq 0°C$	Low/Medium	High	Low
Snow/rainfall > 1mm/h	Low	High	High
Wind speed > 10 m/s	Low	Medium	High
Visibility/fog	Medium	Low	Medium
Ice/snow layer	Medium/High	High	High

Concreting at low temperature is encumber by a risk of damage in concrete structure, especially in early stage of hardening concrete. The main danger for concrete in early state is damaging of structure by freezing water. If the concrete mixture freeze over after initial setting time the water increases volume in concrete pores and destroys the bonding chain in concrete paste. It brings negative results such as lower compressive and tensile strength.

The degree of harmfulness of freezing a concrete mixture increase significantly in stage between initial setting time and end of setting time due to fact that at this curing cycle occur the hardening processes in concrete as shown Figure 1. If the concrete mixture freeze over before initial setting time the hydration process is inhibited or stopped and freeze water is not dangerous for microstructure of paste [3].

The most harmful situation for construction is when the concrete is hard and destructive processes in structure are irreparable. In that case waste of money and time are huge and are strictly connected with demolition of construction. This situation on construction site forced the project managers to analyze the case and the results often shows that prevention as well as prediction could save the construction in future.

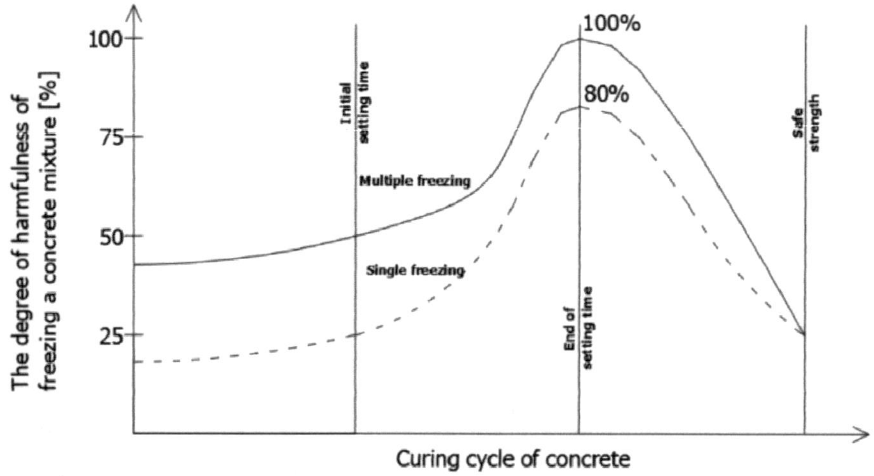

Fig. 1. The degree of harmfulness of freezing a concrete mixture. [3]

In decision making situation, when the quality and quantity of information are low and environment is strongly variable the solution is usage of flexible approach. Implementation of flexibility can be based on two main strategies. First, also called as passive flexibility, refers to action performed before negative scenario occur and is the first component of flexibility to achieve the assumed results of construction process in uncertain environment [4]. The possible actions in passive flexibility are as follow:

- Preparation of proper formula of the concrete mixture, where amount of cement, w/c ratio as well as chemical admixtures and additives are taken into account,
- Using cement with higher amount of cement clinker CEM I and CEM II and also with higher strength class such as 42.5 and 52.5,
- Increasing the proportion of cement by 5-10% in composition of the concrete mixture,
- Using chemical admixtures, which: allow to lowering the freezing point of water in concrete, accelerate the bonding by increasing the initial rate of reactions between cement and water as well as reduce the amount of mixing water,
- Using modern concrete modifiers like PCM (Phase Changed Material).

The second approach, active flexibility, is performed when negative scenario occur. It is a second component of flexibility based on adaptation to changing conditions, especially the basis of active flexibility is monitoring of condition and construction process in progress [5]. Examples of this actions in concreting process are:

- Using specially heated form work to protect the pouring concrete from low temperature,
- Using a heating concrete method when the concrete is heating up by warm air heating, low-pressure steam or electric installation,
- Using covers and foils which insulate the concrete, reducing heat losses,
- Using heating tents, which protect not only against the rain or snow but also against strong wind.

3 Simulation as a trigger for active flexibility option

Modern solutions used in construction process give a possibility to simulate an uncertain cases where discrete events can appear. Simulation is a tool to evaluate a performance of system, existing or proposed, under different configurations of interest and over long periods of real time [6].

Application of simulation in construction process management proceed from limitation of other method such as mathematical programming or queuing theory [7]. The first researcher who proposed usage of simulation in construction equipment choosing was Teicholz in 1963. Ten years later in 1973 Halpin introduced the concept of simulation, bases on cyclic process model. From now the dynamic development of CYCLONE purpose is dated to make an efficient decisions to manage the production systems in construction industry based on simulation. This development process exists to date and some examples of modern software are described at the Table 2.

Table 2. The examples of simulation software and basic features of software [8–12].

Simulation Software	Basic features of software
MicroCyclone	• Fundamental software for Construction Process Simulation • The operations are reduced to simple flowchart what allow to studying the interaction easily
AnyLogic	• Allow to visualize the process in 2D/3D • Different goals of participating could be taken into consideration
VR simulation	• Integrate the MS Access Database, AutoCAD drawings and MS Project Schedules
Stroboscope	• Is based on three-phase activity scanning • Using of graphical network-based representation
SDESA	• Results from the recent construction engineers research • Allow the spatial definition of a construction system in 3D

According to literature the simulation consists of two groups continuous and Discrete-Event Simulation. The difference between continuous simulation and DES is that the first provides the changes in system continuously over time and the second approach assumes the system changes instantaneously. The main aim of Discrete-Events Simulation (DES) is to provide a sufficient solution to modelling processes on building site. Due to simulation of real events engineers can estimate and reduce a time of processes or waste of material and provide security for construction and workers [13].

To make clear an idea of usage a simulation in construction process the further disquisition will be based on concreting in low temperature. Decision making during concreting process, depending on weather condition and weather forecast. Figure 2 presents a role of simulation in concreting process in low temperature.

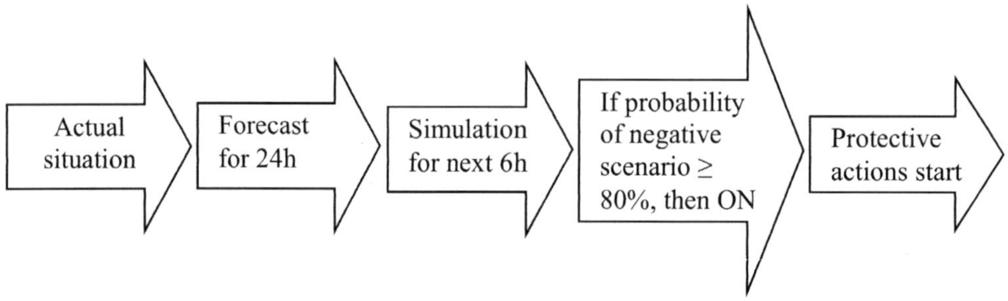

Fig. 2. Role of simulation in process of concreting in low temperature

On the building site, manager has to make a decision about concreting based on weather forecasting which is inaccurate and not stable. This is the reason why the condition on site and weather forecast must be updated. Figure 2 shows the consecutive steps in decision making system. Actual situation is concerned and confronts the forecast for 24h. In this time the delivery and additional task can be performed and the data from monitoring devices are still collected. However, changes that are difficult to predict may occur and therefore a systematic update must be carried out for both: monitored process and weather forecast. The simulation is performed every 6 hours (based on updating process of weather forecast). If the coefficient of risk of damage in concrete structure (C_f) is higher than 80% then the protective action starts. This coefficient is calculated based on simulation according to the following formula:

$$C_f = \frac{\sigma_F}{f_t} \tag{1}$$

where, σ_F - freeze stress,
f_t – tensile strength for maturing concrete.

The main idea of simulation implementation into construction site is presented at Figure 3.

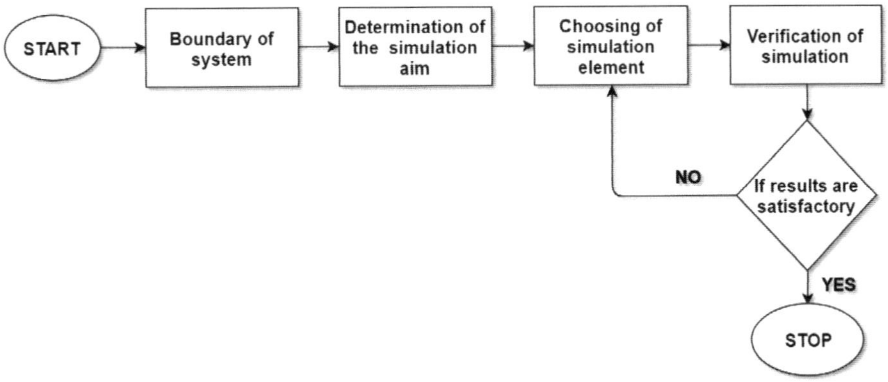

Fig. 3. Schema of simulation process for trigger of active flexibility option

Nowadays there are plenty of simulation software which usage depends on analysed situation. In construction industry are two main groups of simulations. First group refer to designing and simulate phenomena inside elements of construction structure, their compressive and tensile strength, force distribution, and durability over designed time. Examples of software used in this kind of simulation are: Abaqus, HACON and SolidWorks. BIM technology gives possibility to simulate the construction elements exactly from the design documentation, with all enhancements [14]. Dimyadi et al. [13] describe the usage of BIM technology to support simulation of compliant building evacuation.

The second group of software is concerned simulation of the delivery of construction product like ready-mix concrete or works which duration has enormous impact on schedule such as earthwork. The most popular software used in concrete delivery process simulation is MicroCyclone. This type of software allows to choose the number of concrete mixer trucks, reduce the rental costs of trucks as well as maximizing the efficiency of concrete work by eliminating mixer trucks' queuing and downtime [15]. The usage of simulation software in practice will be presented in the following case study.

4 Case study

Presented case study is based on concreting of sealed slab in local petrol station. The thickness of this slab (0.18 m) is relatively small in relation to the width (7 m) and length (10 m). The phenomena in sealed slab during maturing of fresh concrete simulation was performed in HACON program.

The HACON basis on FEM analysis. The theoretical aspect of FEM is based on temperature correlation between finite elements. The final element method is a numerical approach of solving differential equations describing physical problems. The program solve this equation by a finite element formulation and is estimating by a weak formulation with usage of Galerkin method. Simulation process is based on experiments performed in semi-adiabatic conditions (self-heating of concrete) and isothermal conditions (dynamics of increasing the strength of concrete depending on the temperature) in lab. The results from HACON program simulation are presented at Figure 4.

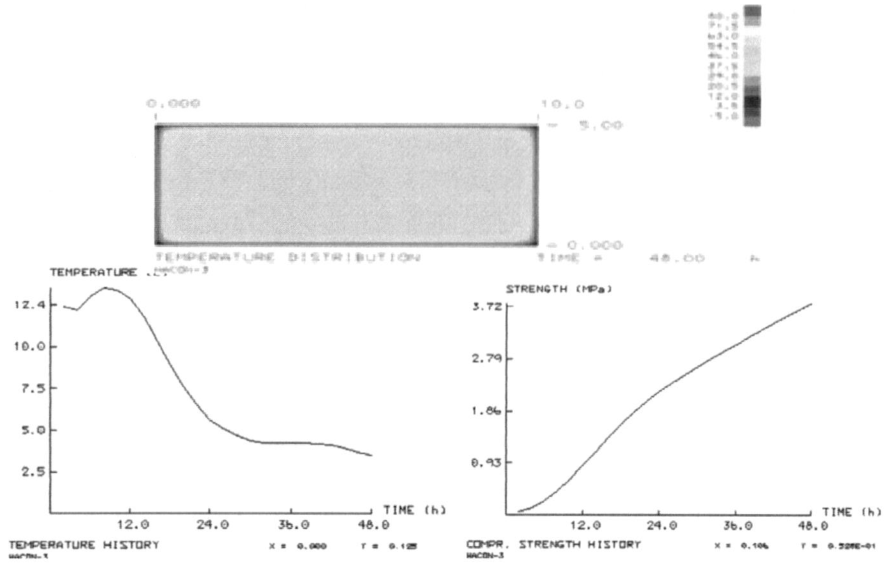

Fig. 4. The result from program HACON- heat and strength distribution in concrete slab

The results from simulation shows that compressive strength after 48 with changes in forecast temperature (with frequent updates), and without this changes, differ and in the traditional situation the unexpected low temperature cause an inhibition in compressive strength growth. Predicted value of compressive strength according to forecasting temperature were planned on 9.52MPa after 48h but in real case the compressive strength was confirmed at 3.8 MPa.

The traditional approach assumes the weather forecast, monitoring the temperature in concrete but rare simulation process (for next 24h or 48h). The changes in temperature within this period of time can be visible and significant, especially in winter condition at night. This is the reason of introducing the new solution with frequent simulation and flexible option. Simulation based on weather forecast and monitoring is frequent (is performed every 6h) what allow to observe the fall in temperature. If the temperature in concrete drops than the additional flexible actions can be done.

5 Conclusions

All theoretical approaches as well as case study discoursed above allows to draw the following conclusions:

1. Traditional approach (no monitoring of construction operation in progress, no simulation of phenomena in concrete element, no updating of weather forecast) for concreting at low temperature is very risky
2. Simulation plays significant role in decision-making process and considers the active simulation option
3. The FEM is more and more popular and commonly used in construction industry
4. Due to usage a BIM technology as well as simulation, processes on building site can be planed and managed more efficiently (what cause smaller wastes of money and time in comparison with traditional approach)

5. The simulation allow to observe the impact of the real changes in weather on heat distribution in concrete slab and allow to trigger protective flexible action.

The further studies can consider the influence of using flexible approach in concreting process depending on changes in weather forecast and more frequent updating actions.

Acknowledges

This work was supported by Institute of Structural Engineering Poznan University of Technology.

References

1. Ballesteros-Pérez, P.; Smith, S. T.; Lloyd-Papworth, J. G.; Cooke, P. Incorporating the effect of weather in construction scheduling and management with sine wave curves: application in the United Kingdom. *Constr. Manag. Econ.*, **85**, 1–17, (2018) doi:10.1080/01446193.2018.1478109.
2. Ballesteros-Pérez, P.; Rojas-Céspedes, Y. A.; Hughes, W.; Kabiri, S.; Pellicer, E.; Mora-Melià, D.; del Campo-Hitschfeld, M. L. Weather-wise: A weather-aware planning tool for improving construction productivity and dealing with claims. *Autom. Constr.*, **84**, 81–95, (2017) doi:10.1016/j.autcon.2017.08.022.
3. M. Pawelska-Mazur Warunki betonownia w obniżonych temperaturach na przykładzie Pomorza. *Przegląd Bud.*, *R.* **82**, *nr*, 25–29, (2011).
4. Andrzejczak, K.; Karlowski, A.; Paslawski, J. O. Flexible management for concrete curing at low temperature based on learning from cases. *34th Int. Symp. Autom. Robot. Constr. ISARC 2017*, 112–119, (2017), doi:10.22260/ISARC2017/0015.
5. Pasławski, J. *Elastyczność w zarządzaniu realizacją procesów budowlanych*; Wydawnictwo Politechniki Poznańskiej, (Poznań 2009)
6. Maria, A. Proceedings of the 1997 Winter Simulation Conference., 7–13, (1997).
7. AbouRizk, S.; Halpin, D.; Mohamed, Y.; Hermann, U. Research in Modeling and Simulation for Improving Construction Engineering Operations. *J. Constr. Eng. Manag.*, **137**, 843–852, (2011) doi:10.1061/(ASCE)CO.1943-7862.0000288.
8. Daniel W. Halpin, L. S. R. *Planning and Analysis of Construction Operations*; John Wiley & Sons, (1992); ISBN 0-471-55510-X.
9. Grigoryev, I. AnyLogic 8 in Three Days. (2018), doi:978 92 4 150215 3.
10. Dawood, N.; Sriprasert, E.; Mallasi, Z.; Hobbs, B. Development of an integrated information resource base for 4D/VR construction processes simulation. *Autom. Constr.*, **12**, 123–131, (2003), doi:10.1016/S0926-5805(02)00045-6.
11. Ioannou, P. G.; Martinez, J. C. Simulation of complex construction processes. In *Proceedings of the 28th conference on Winter simulation - WSC '96*; ACM Press: New York, New York, USA, 1321–1328, (1996).
12. Lu, M.; Wong, L. C. Comparing promodel and sdesa in modeling construciton operations. *Proc. - Winter Simul. Conf.*, 1524–1532, *(2005),* doi:10.1109/WSC.2005.1574420.
13. Kapliński, O. Modeling Construction Processes: a Review of Research and Disserations at the Poznan University of Technology. *Civ. Environ. Eng. Reports*, 85-105, (2005)
14. Dimyadi, J.; Amor, R.; Spearpoint, M.; Compliant, C.; Practice, D. Using BIM to

Support Simulation of Compliant Building Evacuation. *Proc. 11th Eur. Conf. Prod. Process Model. (ECPPM 2016),* 511–518, (2016).

15. Dziadosz, A.; Konczak, A. Decision-making process in the construction industry. *Arch. Civ. Eng.*, **62,** 111–126, (2016).

Delays in construction works in Polish construction industry from the contractors' point of view

Anna Kowalczyk[1], *Wiesław* Meszek[1], *Mariusz* Rejment[2] and *Agnieszka* Dziadosz[1*],

[1]Poznan University of Technology, Piotrowo 5 Street, 60-965 Poznan, Poland
[2]Wroclaw University of Science and Technology, Wybrzeże Wyspiańskiego 27, 50-370 Wroclaw, Poland

Abstract. The construction industry is sensitive to changing random conditions that determine the occurrence of various disturbances during the construction investment process. These disturbances are the source of numerous delays during realization of the construction project. The identification of the reasons for delays and their impact on the progress of construction works allows for the efficient management of the construction project and planning of protective measures, which reduce the risk of their occurrence. The purpose of this article is to identify and then to assess, on the basis of the conducted questionnaire survey, the impact of disturbances on delays in construction works. In addition, the factors were ranked in terms of the degree of impact on the construction project by calculating the rank of the factor.

1 Introduction

Regardless of the scale and complexity of a construction project, the delays are a permanent part in the construction process. By the term "*delay*" we mean the exceeding of the deadline (set out in the contract) for the execution/completion of construction works, which arises as a result of a defined disturbance. The disturbance is a phenomenon that causes: interruption, prolonging the duration of construction works or postponement of execution of construction works, resulting in a delay in the execution time of construction works.

Due to the random nature of disturbances that have a significant impact on the construction process, the problem of delays in construction industry is readily undertaken by lots of scientists [1-5]. The main area of research, presented in the literature, is the identification of delays, the analysis of the basic causes of delays and the attempt to assess their impact on the construction of a building facility, which is often classified as a risk of time or untimely completion of the construction project [1-14]. The authors, in the article, made an attempt to rank factors due to the degree of influence on delays in the execution of construction works based on the preliminary questionnaire survey carried out in May 2018 among people associated with the construction industry, who perform various functions in

* Corresponding author: agnieszka.dziadosz@put.poznan.pl

the construction investment process. The research allowed us to receive a full range of factors causing delays in construction works.

2 Delays in construction works from the contractors' point of view

2.1 Structure of the surveyed respondents

The preliminary questionnaire survey was conducted among contract managers, construction managers, construction engineers, designers and building inspectors in the area of the Greater Poland Voivodeship. The questionnaire surveys forms were sent to 35 randomly selected representatives of construction companies. About 71% of fully completed survey forms were returned. The structure of the surveyed population, due to the work position, is shown in Figure 1.

The largest number of respondents participated in the construction of public buildings (38.3% of respondents). The second place was occupied by single-family and multi-family residential buildings and a collective residence (26.8%). The next item (14.8%) was taken up by cubature and engineering building facilities related to the industrial construction. Less than 13% (i.e. 12.6%) of the surveyed participants took part in the construction of engineering constructions facilities related to the transportation construction. The engineering construction buildings (associated with municipal buildings) constituted 6.6% of the surveyed population. Only 1% of participants participated in the construction of military building facilities. Such a large discrepancy of the constructed building facilities ensures the identification of delays, however, it makes impossible to classify them in relation to the type and type of the constructed building.

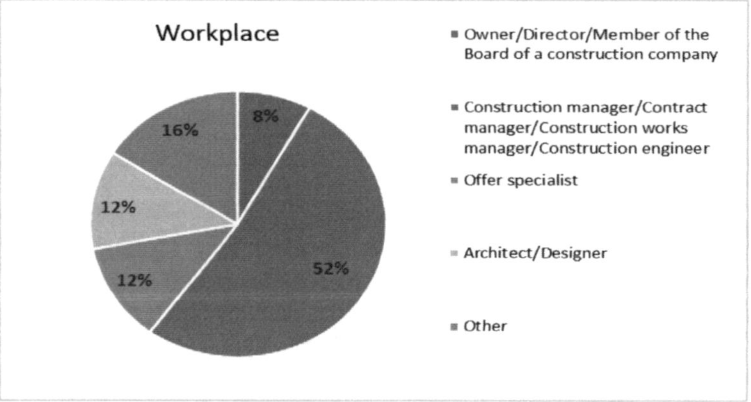

Fig. 1. Structure of the examined population: workplace. Source: authors' data

2.2 Identification and ranking of factors causing delays

The main goal of the research was to identify the most important factors and to determine the extent of their impact on the delays of construction works. The key factors are set together in Table 1. The collected results show that in a given period (i.e. in the year 2018) the lack of workers in the market has had a very big influence on delays. This factor was indicated by 60% of respondents, describing it as "a factor with a very high impact" delaying the completion of construction works. Other factors, that have "a high impact" on delays, are the following:

• too short deadlines imposed by ordering parties (60%),
• deliberate delays in payments (48%),
• difficulties in financing of construction works by a contractor (48%),
• inadequate experience of a contractor/designer (44%),
• delays in the delivery of building materials (36%).

Table 1. List of factors delaying construction works

	Factor	1 – no impact	2 – low impact	3 - medium impact	4 – high impact	5 – very high impact	Rank of factor	
		%	%	%	%	%		
1	Lack of constructive cooperation between an investor (ordering party) and a general contractor	4	8	16	44	28	384	3
2	Weather conditions	4	16	36	32	12	332	7
3	Large number of subcontractors	8	16	40	32	12	348	6
4	Delays in the delivery of building materials	4	12	24	36	24	364	4
5	Application of very modern or very old technologies/construction techniques	8	12	64	16	0	288	8
6	Large construction site	16	28	44	8	4	256	9
7	Deliberate delays in payments	4	0	36	48	12	364	4
8	Incorrect/imprecise construction law	4	24	52	20	0	288	8
9	Too short deadlines imposed by ordering parties	4	0	12	60	24	400	2
10	Staff changes during the construction process	8	8	44	24	16	332	7
11	Lack of workers	4	4	8	24	60	432	1
12	Inadequate experience of a contractor/designer	4	8	28	44	16	360	5
13	Difficulties in financing of construction works by a contractor	4	4	24	48	20	376	4

The following factors had the medium impact according to the respondents:
• application of very modern or very old technologies/construction techniques (64%),

• incorrect/imprecise construction law (52%),
• large construction site (44%),
• staff changes during the construction process (44%),
• weather conditions (36%).

From the perspective of hierarchical ranking of factors, it should be noted that:
• 84% of respondents assessed the impact of the factor "*Lack of workers*" very high, as well as the factor " *Too short deadlines imposed by ordering parties*",
• 50% of respondents rated the factor "*Staff changes during the construction process*", "*Weather conditions*" and "*Application of very modern or very old technologies/construction techniques*" as the factors with a decidedly medium impact or even below,
• according to the respondents, the smallest impact on the delay has the factor "*Large construction site*" and the related with that factor coordination of construction works.

3 Conclusions

In the literature, on this subject, the authors cite many examples of classification of factors causing delays as well as they try to set them into groups. According to the authors of the article, it is worth expanding and refining the nest research, which would allow to determine the impact of the type and specifics of the construction investment, the type of an investor (public, private) and the selected group of participants on the assessment of construction delays. The factors, mentioned in the article, concerned the construction phase of building facility, therefore there are no factor in the respondents' answers, which often appears in the quoted publications, i.e. "*difficulties in obtaining the necessary permits, agreements and opinions*".

The presented factors are an excellent introduction to the risk analysis and attempt to measure it both in terms of time as well as finances. Each of the identified factors has a different impact on the risk level associated with a given construction contract.

Despite the excellent engineers' knowledge, many tools supporting the management of the construction process and numerous studies on the causes of delays in the construction industry, the phenomenon of delays in construction works is still valid. Therefore, it seems necessary to introduce in the next analyzes the dependence of factors on, for example, the type of investor or the specificity of the building facility being constructed.

References

1. J. Drzewiecka, J. Pasławski, Bud. i Inż. Środ., 2, **4**, 475-479 (2011),
2. M. Dytczak, G. Ginda, T. Wojtkiewicz, Bud. i Inż. Środ., 2, **4**, 481-485 (2011)
3. A. Leśniak., E. Plebankiewicz, Zeszyty Naukowe WSOWL, 157, **3**, 332-339 (2010)
4. H. Anysz, A. Zbiciak, Autobusy : technika, eksploatacja, systemy transportowe, R. 14, **3**, 963-972 (2013)
5. A. Kowalczyk, Posnan Univeristy of Technology, master thesis (2018)
6. A. Dziadosz, M. Rejment, Proc. Eng., 122, 258-265 (2015)
7. A. Dziadosz, A. Kończak, Arch. of Civil Eng., 62 (**1**), 111-126 (2016)
8. M. Rejment, A. Dziadosz, Zeszyty Naukowe Politechniki Rzeszowskiej. Budownictwo i Inżynieria Środowiska, 59, **3/III**, 65-72 (2012)
9. M. Gajzler, A. Dziadosz, P. Szymański, Tech. Trans. (1-B), **2**, 107, 71-84 (2010)

10. Zavadska. T. Vilutiene, Z. Turskis, J. Šaparauskas, Arch. of Civil and Mech. Eng., 14 (**1**), 114–121 (2014)
11. O. Kapliński, Proc. Eng., 57, 533–539 (2013)
12. T. Kasprowicz, Arch. of Civil Eng. 62 (**1**), 55-66 (2016)
13. M. Gajzler, K. Zima, Int. J. of Civil Eng., 15, **5**, 641–52 (2017)
14. A. Dziadosz, A. Tomczyk, O. Kapliński, Proc. Eng. 122, 120-128 (2015)

MATEC Web of Conferences **222**, 01009 (2018) https://doi.org/10.1051/mateconf/201822201009
ORSDCE 2018

A comparative analysis of selected models of pedestrian-generated dynamic loads on footbridges – vertical loads

Kinga Marecik[1,*] and *Marek* Pańtak[1]

[1] Cracow University of Technology, Institute of Building Materials and Engineering Structures, Warszawska 24, Cracow 31-155, Poland

Abstract. In the paper the characteristics of selected dynamic load models of vertical forces generated during pedestrian movement (Vertical Ground Reaction Forces - *VGRFs*) and the results of dynamic numerical analyses and dynamic field tests of an exemplary footbridge were presented. Numerical analyses of the footbridge were performed taking into account recommendations of different authors and standards relating to *VGRFs* models as well as *VGR*Fs measured by authors during laboratory tests using force platform. For comparative purposes, the dynamic field tests of the footbridge were carried out. The results and conclusions from comparative analyses of the dynamic responses of the footbridge obtained for various load models and field tests were presented.

1 Introduction

The ways of shaping footbridges have changed over the centuries. Along with the evolution of construction materials not only the size of the cross-sections and structural schemas but also the forms of structures of constructed bridges were changing. Nowadays architects are proposing more and more diverse, complex and abstract shapes of footbridges what have also an impact on its dynamics behavior. Footbridges currently erected, due to their lightness, increased spans length, smaller cross-sections of structural elements and often a complicated form, have become susceptible to dynamic loads generated by pedestrians. The problem of excessive vibration of the footbridges has been clearly noticed at the end of twentieth century (e.g. Millennium Footbridge in London, Léopold Sédar Senghor Footbridge in Paris) and became the subject of interest of a large group of engineers and scientists all over the world. This led to introduction of requirements on the dynamic analyses of footbridges and recommendations related to the assessment of the comfort of use of footbridges due to its vibrations in design standards or design guidelines in different countries. Nonetheless, the problem of footbridge vibration was also noticed earlier and noted in older papers and design standards [1, 2].

Dynamic problems of footbridge can occur when the natural frequency of the structure coincides with the frequency of pedestrian movement. The values of stepping frequencies

* Corresponding author: marecik.kinga@gmail.com

for various form of activity are in the following ranges [3]: 1.4 – 2.4 Hz – walking, 1.9 – 3.3 Hz – running, 1.3 – 3.4 Hz – jumping.

In general, the fundamental frequency of footbridge depends on several coefficients, i.e. structural elements cross-section, materials, static scheme and span length of construction that all have an impact on the mass and the stiffness of structure which are strictly correlated with the formula of the objects natural frequency. According to the research that were carried out on existing footbridges [4] it can be seen that for span length in range 20 – 30 m natural vibration frequencies of the structures are 2.0 Hz to 4.0 Hz; for span length 30 – 50 m natural vibration frequencies are 1.5 Hz to 4.0 Hz; and for span length 50 – 100 m natural vibration frequencies are less or equal to 3 Hz. For the given ranges of spans approximately 60%, 84% and 40% of the structures respectively has a fundamental natural vibration frequency in the frequency range of human movement. These footbridges are exposed to excitation of resonant vibrations. This affects the need to perform the dynamic analyses of the footbridges using appropriate models of pedestrian-generated dynamic loads.

2 Models of pedestrian-generated vertical dynamic loads

Pedestrian-generated vertical dynamic loads (Vertical Ground Reaction Forces (*VGRF*s)) are described by various authors in form of different mathematical models. To study the accuracy and correctness of some of these load models the three load models representing the *VGRF*s generated during human walking (Eqs. (1) - (3)) and the *VGRF* measured by means of force platform during laboratory tests of walking person were analyzed.

The load model No. 1 was adopted in accordance with BS-5400-2:1978 [1] as a pedestrian load model described by means of a formula:

$$F(t) = 180\sin(2\pi \cdot f_0 \cdot t) \tag{1}$$

where: f_0 – natural vibration frequency [Hz], t – time steep [s].

The model No. 2 was taken from NA-BS-EN 1991-2:2003 [5]. The function representing *VGRF*s is much more complex than given by the equation (1) and depends on several coefficients depending on the type of pedestrian traffic, footbridge location, footbridge natural vibration frequency and its mode shape, value of the logarithmic damping ration and is given as follows:

$$F(t) = F_0 \cdot k(f_0)\sqrt{1+\gamma(N-1)} \cdot \sin(2\pi \cdot f_0 \cdot t) \tag{2}$$

where: F_0 – reference amplitude [N], $k(f_0)$ – combined factor, γ – reduction factor for unsynchronized action of pedestrians, N – the number of pedestrian in the group, f_0 - natural vibration frequency [Hz], t – time steep [s].

The load model No. 3 was the *VGRF* pattern generated by one foot [6]. It represents more adequate values of force generated during pedestrian movement by each foot and shows approximation of the accurate physical situation of generated vertical force during walking. It is described by the formula:

$$F(t) = G\sum_{i=1}^{5} A_i \sin\left(\frac{\pi \cdot i \cdot t}{t_c}\right) \tag{3}$$

where: G – body weight of walking person [N], t_c – time of the foot contact with the ground [s] ($t_c \approx 1.32\ T_w$, T_w – period of walk, T_w=1/f_0), f_0 - natural vibration frequency, t – time steep [s], A_i – amplitude of the i-th harmonics, i – the number of the harmonics, i= 1 - 5.

The last load model (Lab Test *VGRF*) used in analyses come from laboratory tests and represents the real values of dynamic impact of human on the ground. The Lab Test *VGRF* was obtained during gait tests of one person (body weight G = 840 N) carried out using the force platform Zebris FDM-1.5.

All *VGRF*s functions representing by loads models No. 1 – 3 and Lab Test *VGRF* are presented in Fig. 1. In all cases, walking frequency was assumed as equal to 2.38 Hz. The formulas components in equations (2) and (3) were chosen based on the parameters of selected steel footbridge in Osjaków briefly characterized in Chapter 3.

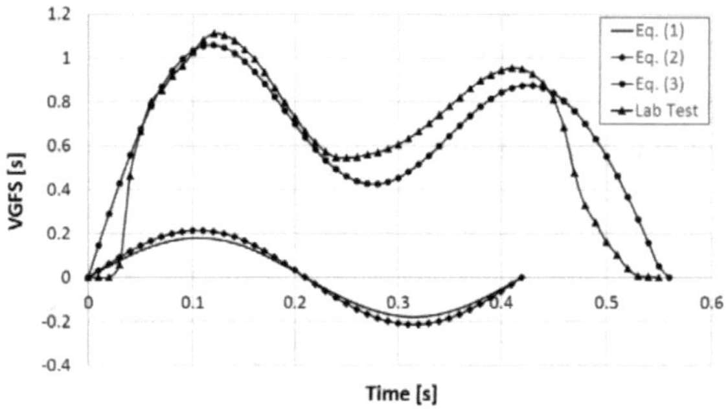

In equations (2) and (3) the following values were assumed:

F_0 = 280 N	G = 840 N
$k(f_0)$ = 0.72	A_1 = 1.047
γ = 0.13	A_2 = 0.092
N = 2	A_3 = 0.544
Class A	A_4 = 0.0533
	A_5 = 0.00176

Fig. 1. The comparison of the values of the *VGRF*s load models used during footbridge analysis.

3 Dynamic analysis of the footbridge

To perform comparative analyses of all load models characterized in Chapter 2, a numerical 3D model of a selected steel footbridge characterized by high dynamic susceptibility was made.

3.1 Characteristic of the footbridge

Footbridge in Osjaków is a steel footbridge with a span length L = 50.00 m. The general view and the cross-section of the footbridge are shown in Fig 2.

Fig. 2. General view and cross-section of footbridge in Osjaków.

A main girder of the footbridge is designed as a spatial truss contains 13 pyramidal parts which are spaced in intervals of 3.85 m. Upper and lower chord of the truss consists of tubular section Ø323.9 mm and have variable thicknesses along with the length of the span length. The upper and the lower chord are connected by a three-dimensional system of diagonals, posts and vertical beams. The deck of the footbridge (8.0 mm steel sheet) rests on the longitudinal beams HEB100 supported on cross-beams HEB140.

3.2 Dynamic characteristic of the footbridge

The dynamic characteristic of the footbridge was obtained from the numerical analysis and verified during the dynamic field tests of the footbridge. A fundamental natural vibration frequency of the footbridge is f_0 = 2.38 Hz (vertical mode shape, Fig. 3). This value is in the stepping frequency range presented in Chapter 1 corresponding to fast walking or slow running (jogging). Further dynamic analyses of the footbridge were made taking into account the *VGRFs* generated during walking.

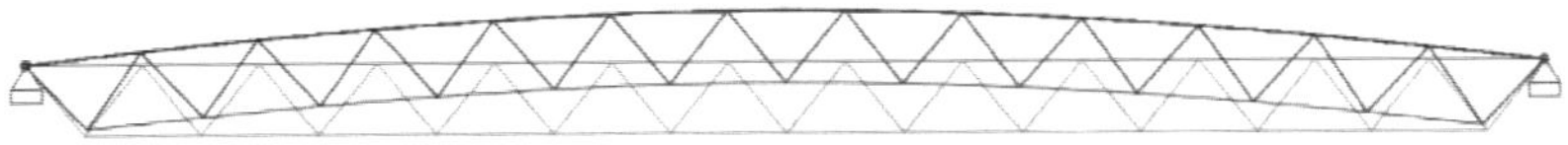

Fig. 3. The fundamental mode shape of the footbridge in Osjaków f_0 = 2.38 Hz.

In analyses the vibration dumping parameter - the logarithmic decrement δ = 0.0076 (corresponding damping ratio ζ = 0.0012) determined on the basis of the footbridge field tests results obtained during resonant vibration excitation by one walking person was used.

3.3 Comparison of acceleration obtained for different load models

Numerical dynamic analyses of the footbridge were made using the dynamic load models No. 1- 3 and the Lab Test *VGRF* described in Chapter 2. As a results the values of vibration accelerations of structure ware obtained. The maximum values of acceleration received from dynamic analyses are shown in Tab. 1. As a comparative value the value of maximum acceleration obtained during the in-situ dynamic tests of the structure in the case of resonant excitation of vibrations by one walking person was used (last row in Tab. 1).

Table 1. Vibration acceleration obtained during numerical analysis and in situ tests.

Load model	Mass [N]	Vibration acceleration [m/s^2]
Eq. (1)	-	0.76
Eq. (2)	-	1.15
Eq. (3)	840 N	2.28
Lab Test *VGRF*	840 N	2.09
In-Situ Test	840 N	1.95

In Fig. 4 the dynamic response of the footbridge as a function of time obtained for models No. 1 and 2, Lab Test *VGRF* and In-Situ Test were presented.

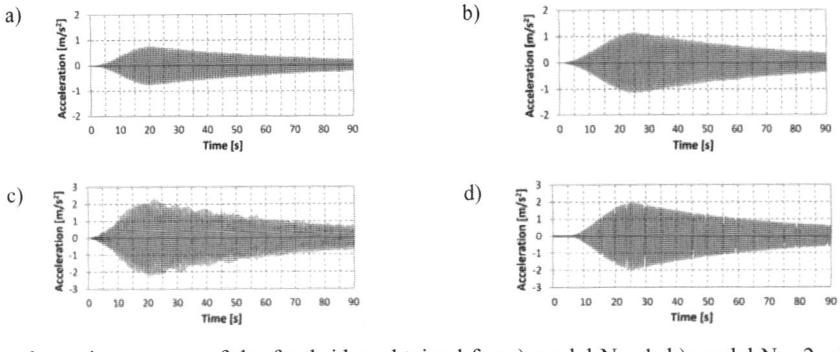

Fig. 4. The dynamic response of the footbridge obtained for a) model No. 1, b) model No. 2, c) Lab Test *VGRF*, d) In-Situ Test.

4 Conclusions

The analysis of the results allows to formulate some important conclusions:

1. The force amplitudes in model No. 1 and 2 are $F_{0,(1)}$ = 180 N and $F_{0,(2)}$ = 214 N respectively. These values differ insignificantly. For number of pedestrian N = 1 and $k(f_0)$ = 0.643 force amplitude $F_{0,(2)}$ = 180 N. This indicates a similar method of calibration of the models No. 1 and 2 for the case of walking of one person.
2. The values of vibration accelerations determined using models No. 1 and 2 are underestimated. Models No. 1 and 2 are inappropriate to determine the dynamic response of the footbridges induced by one person walking in resonance with natural vibration frequency of the structure. The vibration amplitudes calculated using models No. 1 and 2 are respectively 2.57 and 1.69 times smaller than In-Situ Tests result. It can mean that the models No. 1 and 2 are calibrated to determine the dynamic response of the structure induced by walking person unsynchronized with the natural vibration frequency of the structure. To confirm this observation, further analyses and dynamic in-situ tests are necessary. Uncritical use of models No. 1 and 2 can lead to incorrect estimation of vibration amplitudes.
3. The convergence of vibration acceleration value estimated using model No. 2 with In-Situ Test results is reached after increasing the amplitude $F_{0,(2)}$ of about 62% to the value of 346 N ($F'_{0,(2)} = 1.62F_{0,(2)}$). This correspond to the assumption, according to [5], pedestrian number N = 16 in model (2). To achieve the convergence of results for the model No 1, the amplitude $F_{0,(1)}$ should be multiplied by the factor 2.50 and increased to the value of 450 N ($F'_{0,(1)} = 2.5F_{0,(1)}$). The differences in new force amplitudes in models No. 1 and 2 resulting from differences in the walking speed defined for the models ($0.9f_0$ = 2.14 m/s and 1.7 m/s respectively), which affect the number of the *VGRF*s included in 3D computational model of the footbridge.
4. The Lab Test *VGRF* was a force model that gave the value of vibration acceleration closest to In-Situ Test result (with only 7% of discrepancy). This result can be assumed as a very accurate estimation of vibration amplitudes of the footbridge.
5. The values of vibration accelerations determined using model No. 3 exceed the In-Situ Tests result by 17%. This result can be considered as a good and conservative estimation of vibration amplitude. Moreover, using the models No. 3 the *VGRF*s patterns (Fig. 1) and the values of vibration accelerations (Tab. 1) are similar to those obtained using the Lab Test *VGRF*. This proves correctness of the *VGRF*s pattern determination using equation (3). Model No. 3 can be considered correct and can be used in the design process since it gives good and conservative estimation of the vibration amplitudes during resonance excitation.

References

1. J. Blanchard, B.L. Davies, J.W. Smith, *Proc. DOE and DOT TRRL Symp. on Dynamic Behaviour of Bridges*, (1977)
2. BS-5400-2:1978, *Steel, concrete and composite bridges. Part 2: Specification for loads*, (BSI, 1978)
3. H. Bachmann, W. Ammann, F. Deischl et al. *Vibration Problems in Structures: Practical Guidelines*, (Birkäuser Verlag, Basel, Boston, Berlin, 1995)
4. O. Fisher, M. Pirner, *Proc. 2nd European-African Conf. on Wind Engineering* (1997)
5. NA-BS-EN 1991-2:2003, *UK National Annex to Eurocode 1: Actions on structures – Part 2: Traffic loads on bridge*, (BSI, 2008)
6. Q. Li, J. Fan et al., J. Sound Vibr. **329**, 13 (2010)

Estimation of certain parameters of Black-Scholes model in analysing effectiveness of development investments

Wiesław Meszek[1] and *Agnieszka* Dziadosz[1*]

[1]Poznan University of Technology, Piotrowo 5 Street, 60-965 Poznan, Poland

Abstract. The option pricing theory has wide applicability in corporate finance, but it is also increasingly used to analyze the effectiveness of non-financial (material) investments. In traditional investment analysis, a project or a new investment should be accepted only if the returns on the project exceed the hurdle rate; in the context of cash flows and discount rates, this translates into projects with positive net present values (NPV). There is no doubt that it does not take full account of the numerous options that usually relate to developer investment. However, in many cases, the valuation of real options is more difficult than the valuation of options for financial assets. In this paper, we will analyze one of the options, which isembedded in capital budgeting projects - the option to delay a project, especially when a the company has exclusive rights to the project. The value of the option is largely derived from the variance in cash flows – the higher the variance, the higher the value of the project delay option. The variance in the present value of cash flows from the project can be estimated in different ways, however, in the case of non-financial investment projects, these methods are very limited. We are analyzing the possibility of estimating this volatility, taking into account the fact that the forecasted cash flows may show varying volatility in individual years. The paper shows, that by using a probability-based valuation model (using the Crystal Ball techniques) it is possible to incorporate uncertainty into the analysis. The method of presented volatility estimation can be applied by taking into account the randomness of many input data to the project.

1 Introduction

Projects are typically analyzed based upon their expected cash flows and discount rates at the time of the analysis; the net present value calculated on that basis is a measure of its value and acceptability at that time. Assuming that a project requires an initial up-front investment of X, and that the present value of expected cash inflows calculated right now is V, the net present value NPV of this project is the difference between the two:

$$NPV = V - X \tag{1}$$

* Corresponding author: agnieszka.dziadosz@put.poznan.pl

The company's decision rule on this project can be summarized as follows:

- if V > X → take the project: Project has positive net present value,
- if V < X → do not take the project: Project has negative net present value.

If the expected cash flows on the project were known with certainty and were not expected to change, there would be no need to adopt an option pricing framework, since there would be no value to the option. Expected cash flows and discount rates change over time and so does the net present value. Thus, a project that has a negative net present value now may have a positive net present value in the future (the project may still be a good project if the company can wait. waits).

The changes in the project's value over time (the present value of the cash inflows may change over that time, because of changes in either the cash flows or the discount rate) give it the characteristics of a call option.

In the article, we consider the situation that the company has exclusive rights to a particular project for the next one year, so we are considering the applying the option pricing theory to valuing the option to delay. The uncertainty of expected cash flows on the project should be viewed as the reason for why the project delay option has value.

The inputs needed to valuing this option are the same as those needed for any option; we need the value of the underlying asset, the variance in that value, the time to expiration on the option, the strike price, the riskless rate and the equivalent of the dividend yield (cost of delay). In the case of real options, the underlying asset is the project itself. The current value of this asset is the present value of expected cash flows from initiating the project now.

Assuming that the cost of delay is zero, the value of a call option for a non-dividend-paying underlying stock in terms of the Black-Scholes parameters is:

$$C = V \cdot N(d_1) - X \cdot e^{-rt} \cdot N(d_2) \tag{2}$$

where:

C – Call Value,
V – current value of the underlying asset (present value of expected cash inflows from Project),
X – strike price of the option (an initial up-front investment)
t – life to expiration of the option (period of exclusive rights to Project),
σ – variance (volatility) in the value of the underlying asset (the variance/volatility in the present value of cash flows from Project),
r – riskless interest rate corresponding to the life of the option,
N(d1) – cumulative distribution of the standard normal distribution for the argument d1,
N(d2) – cumulative distribution of the standard normal distribution for the argument d2;

$$d_1 = \frac{\ln\left(\frac{V}{X}\right) + t \cdot \left(r + \frac{\sigma^2}{2}\right)}{\sigma \cdot \sqrt{t}} \tag{3}$$

$$d_2 = d_1 - \sigma \cdot \sqrt{t} \tag{4}$$

Calculating the value of the call option, and then determining the expanded value of NPV, i.e. ENPV, allows the final conclusion about the effectiveness of a given project [2].

$$ENPV = NPV + C \tag{5}$$

where:
ENPV – expanded NPV,
C – call option.

2 The variance in the present value of cash flows from the project and the possibility of estimating it

The value of the option is largely derived from the variance in cash flows – the higher the variance, the higher the value of the project delay option. Thus, the value of an option to do a project in a stable business will be less than the value of one in an environment where technology, competition and markets are all changing rapidly.

Assuming (for the analyzed case study) that:
Project will generate 100 000 EUR in after tax cash flows for the next 5 years, discount rate is 10% (r=10%), we obtain:

$$V = \sum_{i=1}^{5} \frac{CF_i}{(1+r)^i} = \sum_{i=1}^{5} \frac{100}{(1+0{,}1)^i} = 379{,}1 \tag{6}$$

Continue to assume, that an initial up-front investment would be equal to the present value of expected cash inflows from Project (X=379,1), NPV of this Project would be zero (NPV=0).

Calculating the exclusive rights to this Project using the Black-Scholes formula for different variance parameters (time to expiration is 1 year, riskless rate is 3,5%) we obtain the results shown in the figure below.

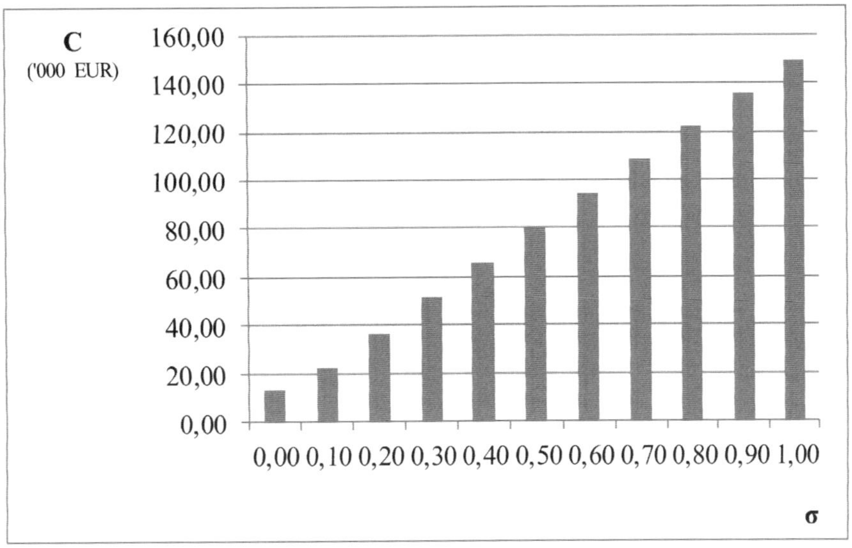

Fig. 1. The value of the call option (the option to Delay a Project), depending on variance in the present value of cash flows from Project

The above chart confirms the hypothesis that the value of the Call option depends very much on the volatility in the underlying project, which confirms that the volatility coefficient estimate is a very important element of the model.

The variance in the present value of cash flows from the project can be generally estimated in one of three ways [1]:

- if similar projects have been introduced in the past, the variance in the cash flows from those projects can be used as an estimate,
- probabilities can be assigned to various market scenarios, cash flows estimated under each scenario and the variance estimated across present values,
- the variance in company value of companies involved in the same business (as the project being considered) can be used as an estimate of the variance.

In the case of individual development investments, however, the above methods are useful to a very limited extent.

3 Case study

In our analysis, the uncertainty has been incorporated into the explicit model of NPV. This is done by recognizing that the input variables (in our case, projected cash flow and discount rate) are uncertain and will have a probability distribution pertaining to each of them. Thus by utilizing a probability-based valuation model (using the Crystal Ball technique) it is possible to incorporate uncertainty into the analysis.

The Crystal Ball technique allows incorporating uncertainty into the analysis in a relatively simple form. It is a simulation model (using the Monte Carlo technique) that, instead of taking one defined set of input figures and producing a single point answer (value), carries out multiple calculations through an iterative re-sampling process. Each simulation chooses an input variable from within the probability distribution chosen for each variable and marries these with other randomly chosen inputs to produce a value. The output is expressed as a range of possible values.

In this valuation triangular probability distributions indicated in the literature as appropriate to account for uncertainty in valuation processes [3,4] were used. The manner in which the individual random variables are recognized is indicated below.

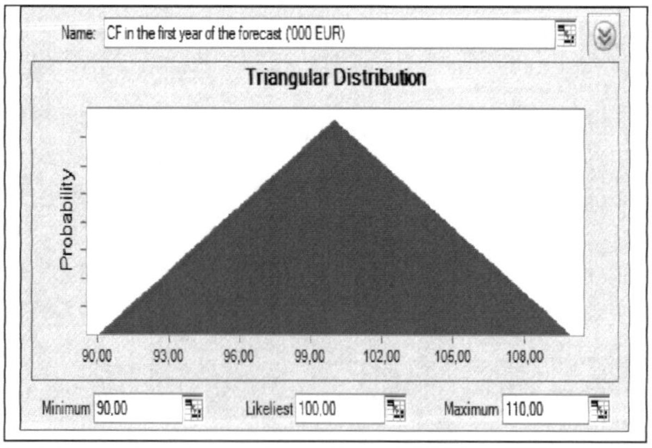

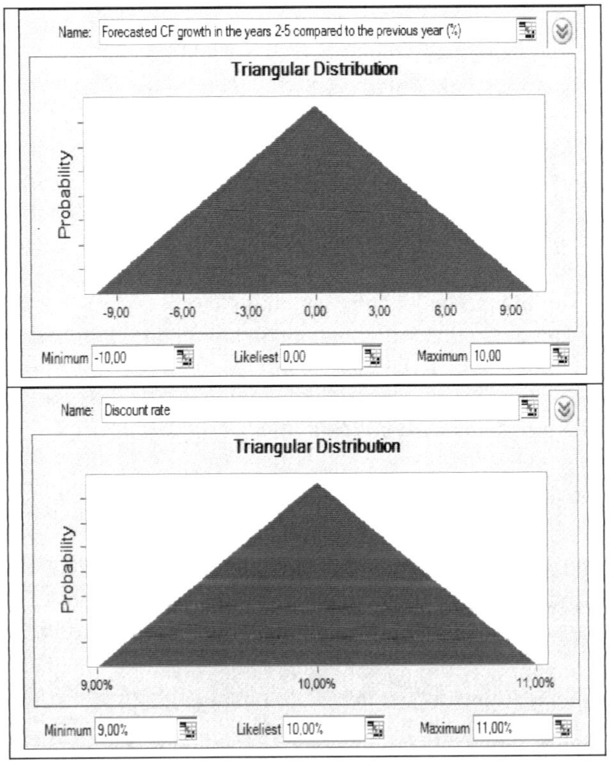

Fig. 2. The form of individual random variables

The results of using the Crystal Ball program were illustrated by the graph below.

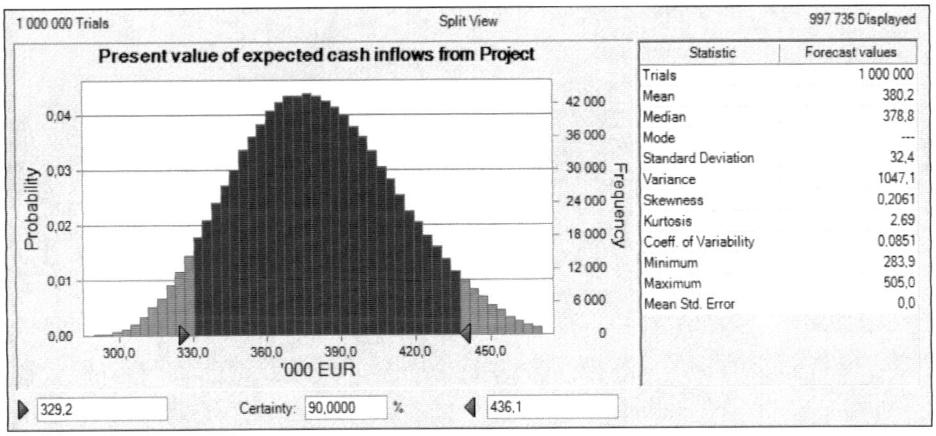

Fig. 3. Simulation results

Using the coefficient of variability given above (s=0.0851, which means variance σ=0,0072), the value of the call option (the option to Delay a Project), calculated based on the formulas (2), (3) and (4), is 20,2 thousand EUR.

In this case it meant that also the expanded NPV of the investment is:

ENPV = NPV+C = 20,2 thousand EUR

In many cases, this is the only way to objectivization of the process of analyzing the effectiveness of development investments using real options.

4 Conclusion

The option pricing theory is widely used in corporate finance, but is also increasingly used to analyze the effectiveness of non-financial investments, including property development investments.

The value of the Call option depends very much on the volatility in the underlying project; in many cases this refers to the variability of the cash flows from the project. Estimation of the coefficient of variation is therefore an important element of the calculation model.

The variance in the present value of cash flows from the project can be estimated in different ways, however, in the case of non-financial investment projects, these methods are very limited.

The simulation process is a universal method of estimation of the coefficient of variation, which is shown in the analyzed example.

References

1. A. Damodaran, (NYU Working Paper No. S-DRP-05-02. Available at SSRN: https://ssrn.com/abstract=1295849; 2005)
2. J. Mizerka, J. Surma, , Rok LXII, 1(2000)
3. N. French, L. Gabrielli, J. of Prop. Invest. & Finan, Vol. 22 No. 6, 484-500 (2004)
4. N. French, L. Gabrielli, J. of Prop. Invest. & Finan., Vol. 23 No. 1, 76-89 (2005)

Adaptability in buildings – housing context – literature review

Roman Milwicz[1,*], *Jerzy* Pasławski[1]

[1]PUT, Civil ad Environmental Engineering, Piotrowo 5, 60-101 Poznań, Poland

Abstract. Dynamic development of technology that influences social lifestyles has great impact over the way of understanding the house and thus the way of designing this special space. The buildings have to fulfill the sustainability conditions and a lot has been done to decrease impact of building on environment. Among many ways one is the flexibility, which strongly influences economical, social and environmental aspects. Authors tried to collect most significant papers and monographs that treat the flexibility and adaptability and thus show its importance in civil engineering especially in housing sector. In the conclusions Authors shears the ideas of investigations within housing adaptability filed.

1 Introduction

The building is not a static object; it remains unfinished until the day of its demolition. This phenomenon had been known since prehistoric times. The buildings is constructed to be useful but needs are not permanent, they are changing in various time horizon from phases of a day, seasons, years. Sometimes need change with time in other case with the change of user or development of technology. People always tried to use the building, as long as possible that is why a lot of ways for adaptation has been introduced during the history of human beings. Thoughts related to adaptability and the time of they appearance is presented in the fig.1. Contemporary civil engineering and architecture have important role to develop, evaluate and disseminate the knowledge connected with adaptability with indeed plays considerable role in sustainable development. Thus the authors standing at the threshold of investigations concerning adaptability in single family housing, would like to present the literature review and indicate the milestones within the field of study.

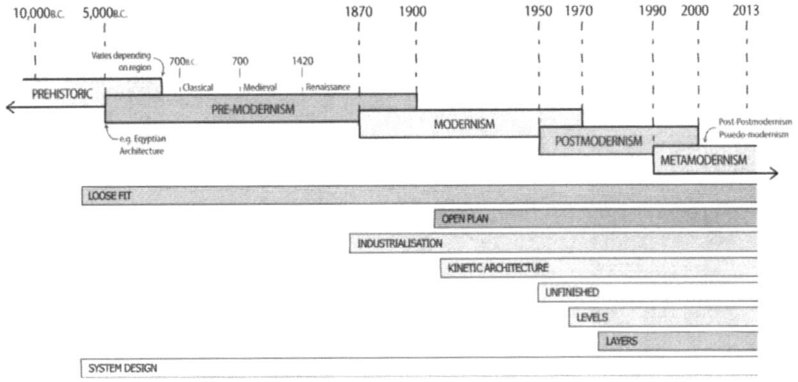

Fig. 1. Development of adaptability [1]

* Corresponding author: roman.milwicz@put.poznan.pl

2 Historical overview

At the beginning people look for shelter in the natural formations which were mostly caves, because of nomadic lifestyle they had to adapt and create the shelter themselves – inventing primitive huts and tents. It was discovered quite early that it is more practical to create bigger space in advance than rebuild the hideaway. This way of adaptability was major for long time up to the Industrial Revolution. Bigger precision and possibility of mass production open the way to build cheaper and faster. First well known building was Cristal Palace erected in 1851 from precast elements. Bold plans of Le Corbusier and his rules gave new quality of buildings. His open plan [2] presented on domino house in 1914 allowed big adaptable performance. It split the bearing structure and other elements. Thank to this solution the space can be created freely by the user. Paralelly the lack of space force designers to create the building that can change quickly and repeatedly the space by e.g. moving portions, such houses are called convertible [3]. Recent 50 years bring a lot of knowledge, which is presented, in next paragraph.

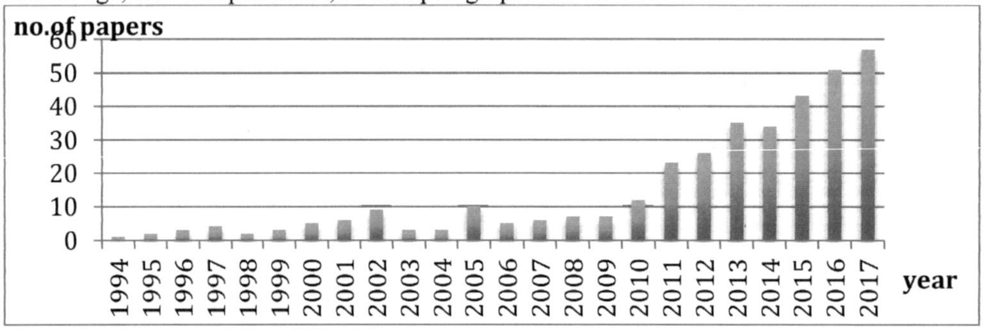

Fig. 2. Development of adaptability [4]

3 Designing for adaptability – general review

Adaptability becomes more and more recognisable term; one can observe substantial development in number of publication touching the adaptability in building context fig.2. "If the building does not support (technologically and technically) change and reuse you have only the illusion of sustainable construction [5] - this sentence shows how important is the adaptability in sustainable development. Some attributes of adaptability are as follows: in the housing heating context the most important seems to be the ability to change volume, function or performance [6], ease of response to the changed conditions [7] presence of less common but more dramatic changes [8]. Adaptability of buildings means to remain ready for change in order to respond the needs or reduce the mismatch [9]. Schmidt III and Austin had done great work; Authors would like to show the strands of designing for adaptability (fig.3.), which explain the progress of adaptability in architecture. [1] The diversity of adaptation methods may give a blurred image of what this term means and how to implement this feature, thus the framework of main types is presented in fig.4.

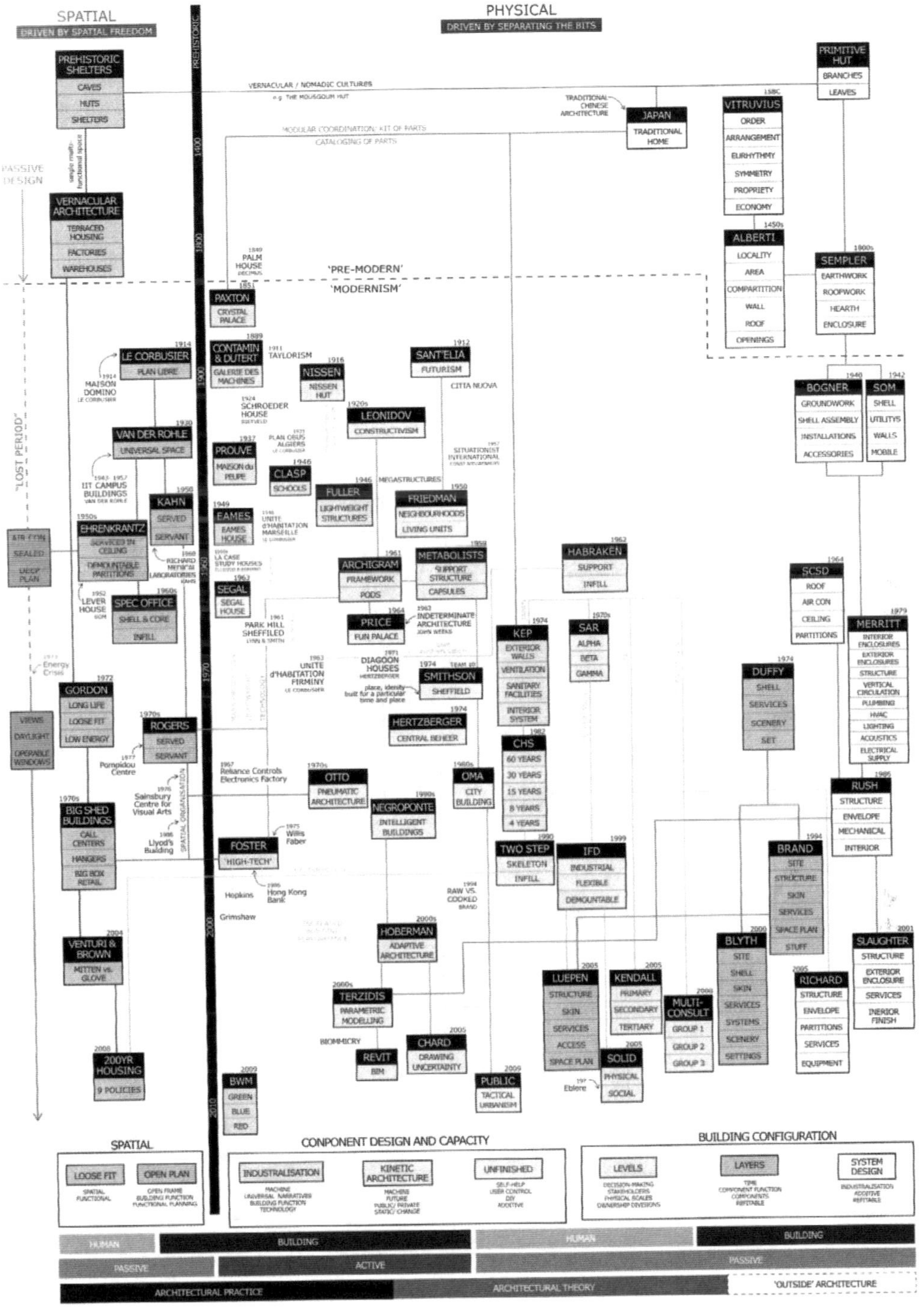

Fig. 3. Strands of designing for adaptability [1]

The implementation of adaptability needs the understanding the components of the building their cooperation and lifespan. The building decomposition is crucial in adaptable buildings the idea of shearing layers was introduced by Duffy and Henney in 1989 and developed by Brand in 1994 and later by Blyth in 2000. The building element are subdivided into sharing layers with different lifespan expectancy (fig.5.):

- Site: the land where the building is erected
- Structure: load bearing elements e.g. foundations, beams columns walls
- Skin: protection from external factors: cladding – walls and roof
- Service: Working guts of the building e.g. electrical, plumbing, HVAC
- Space plan: interior layout
- Stuff: furniture [10]

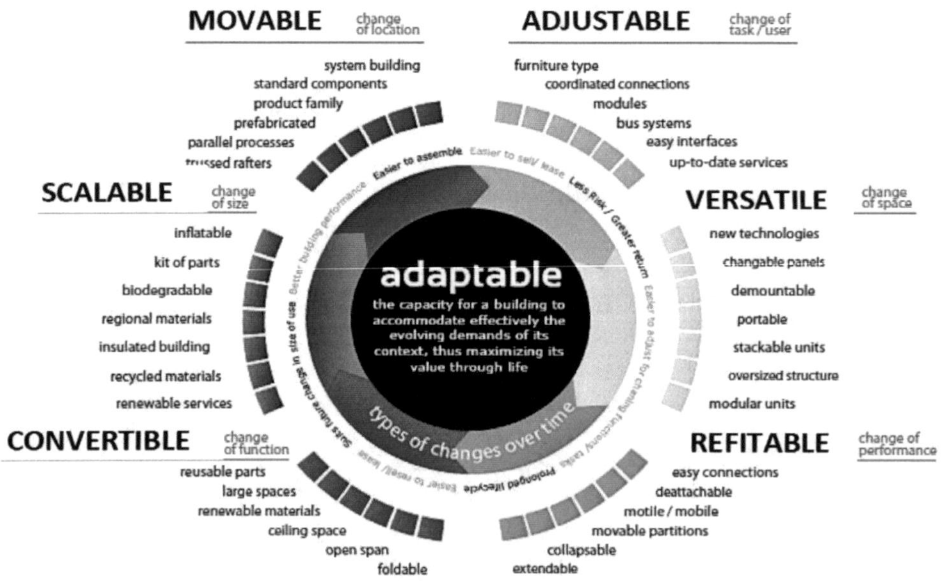

Fig. 4. Adaptable framework [11]

The most important lesson learned form this division is that each sharing layer has different life expectancy and needs to be replace or repair during the life time in various time thus it is essential to design building in such a way to make this process easy and cost effective; the simplest way is not to mix functions of single element among different layers. In other words The 'theory of layers allows the components of the building to be broken down into packages of same or similar life expectance so that a whole package might be conveniently deconstructed from the building for replacement, recycling and/or reuse elsewhere' [12].

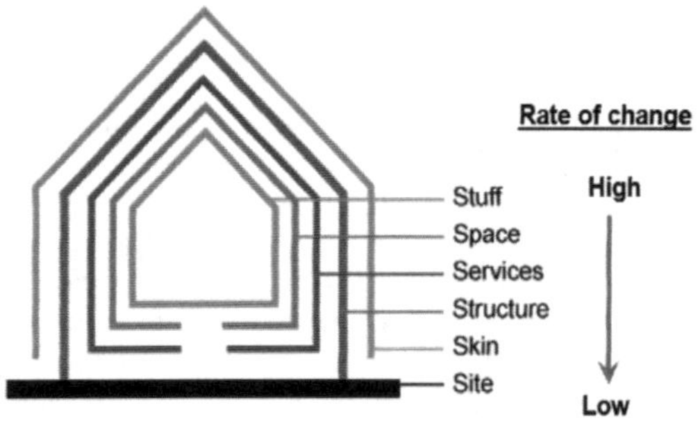

Shearing layer	Life expectancy
Site	Eternal
Structure	30 – 300 years
Skin	20+ years
Services	7 – 20 years
Space plan	3 years
Stuff	Under 3 years

Fig.5. Sharing layers with life expectancies.

There is the connection between sharing layers and adaption types presented in fig.4. One can observe the difference in frequency of changes among different layers. More frequent changes are with little importance and take place in interior and can be done by user, on the other hand bigger changes are less frequent and involve designers within the process mainly because they interfere in external and load bearing components. Next step of investigations might be adaptive reuse potential introduced by Langstom [13]. This is the tool for measuring the potential of existing buildings to adopt for future potential change. Introduced model [14] ranks the potential of adaptation and can be implemented in all types of buildings all over the world. The ARP model take into account the age predicted lifespan and rate of obsolesce. The better score the higher potential for success.

Table 1. The influence of shearing layers in different adaptable strategies.

Strategy	B-E Scale	Speed of change	Shearing layers					
			Stuff	Space	Services	Skin	Structure	Site
Adjustable	Components	Daily/monthly	•					
Versatile	Components	Daily/monthly	•	•				
Refitable	Components	7 years		•	•	•		
Convertible	Building	15 years		•	•	•		
Scalable	Building	15 years		•	•	•	•	
Moveable	Building	30 years					•	•

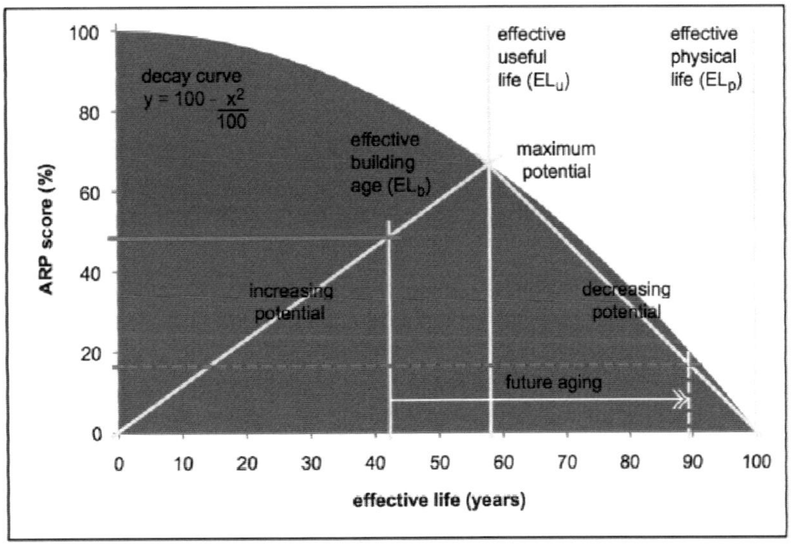

$$\text{ARP (increasing)} = \frac{\left(100 - \frac{EL_u^2}{100}\right) \cdot EL_b}{EL_u} \qquad \text{ARP (decreasing)} = \frac{\left(100 - \frac{EL_u^2}{100}\right) \cdot (100 - EL_b)}{100 - EL_u}$$

Fig.5. Adaptive reuse potential model [13]

4 Adaptability in housing

The last topic that Authors would like to touch is the adaptability in housing sector. The main reason is that the house or flat is really complex type of building mainly because of diversity of activities. Estaji neatly describes the phenomenon [15]. In his works he describes the concept of "Ideal House" that covers all human needs. But the problem lays in rapid change of needs. The paper presents complexity of the house task, with have to answer human needs, on the one hand 12 phases of life (long time horizon -years and decades) fig.6., on the other hand type of activity (short time horizon - hours and days) fig.7. What is important and have been mentioned by Milwicz [16] not only needs are changing but also financial possibilities fig. Such relationship makes designing problem even more complicated.

Twelve Phases of Human Development

Phase	Name of phase	Ages
1	Pre-natal	-9mo.-0
2	Breast dependence	0-6 mo.
3	Infant	7-15 mo.
4	Toddler	16-29 mo.
5	Preschool	30mo.-5y
6	School age	6-12
7	Adolescence	13-18
8	Young adulthood	19-25
9	Middle adulthood	26-40
10	Real adulthood	41-60
11	Early old age	61-75
12	Old age	76-100

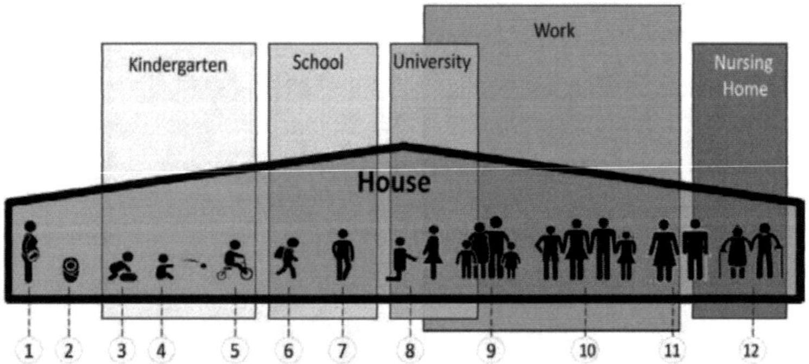

Fig.6. Comparison of the coverage of human development phases by different buildings [15]

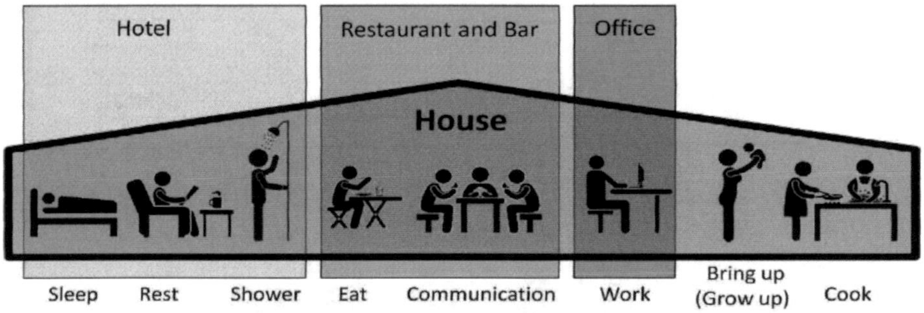

Fig.7. Human activities [15]

In spite of great importance of adaptability in housing sector only a few countries conduct the investigation and introduced the appropriate housing programme namely in UK by Habraken [16] and Schneider, T., & Till, J. [17] , Germany by Huber [18] and Schittich [19] and in Japan [20] in this country the project has started in '70 and results influenced the construction codes extending life cycle of building.

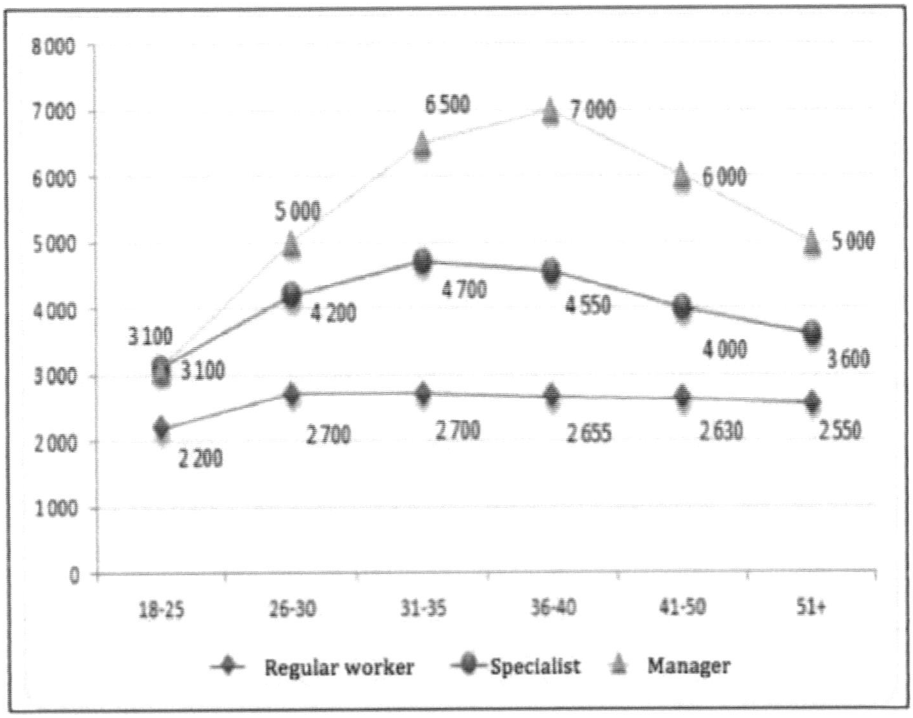

Fig. 8. Salary [PLN] vs. age of worker [years]

5 Summary

Lively interest in adaptation issues will surely will give fruits and help to construct more sustainable buildings. The investigations shows that it is not a single field involved in the case, it a combination o actions such as architecture, civil engineering, environmental engineering, psychology, sociology and many others. The investigations also give tangible results in the field or restoration the example could be paper of Jensen [21]. Nowadays construction reached the satisfying level of energy efficiency and material optimization, now it is time to make flexible buildings than can easily adapt to more and more rapid changes.

From the literature reader can derive following conclusions:

1. There is high need to implement adaptability in buildings
2. Adaptability can be divided into passive and active
3. Main factor of adaptability are: height of rooms, separate construction, avoidance of bearing walls inside the plan, modularity.

Authors see the need to investigate the concept in single-family housing. The concept is to check what kind of adaptability can bring bigger profits. They intend to compare regular house with modular adjustable (active adaptability) and house with the possibility to adapt the attic for future needs (passive adaptability). The special place is dedicated movable modules that can be added or disconnected to the building in any time. Next article will explain the concept and show the possibilities of applying. According to Authors the concept can increase the housing affordability.

References

1. R.Schmidt III, S. Austin, Adaptable Architecture,Theory and practice, Routledge Taylor & Francis Group, NY (2016)
2. Le Corbusier, Vers une architecture,(1923)
3. A. Friedman, The adaptable house: Designing homes for change. McGraw-Hill. New York (2002)
4. Science direct statistics (access 3.07.2018)
5. W. Croxton, Disassembly and Deconstruction, Architectural Record, (2003)
6. J. Douglas, Building adaptation. 2nd ed. Great Britain, Elsevier Ltd. (2006)
7. R. Kronenburg, Flexibile Architecture that responds to change. Laurence King Publishers. London (2007)
8. A. Leaman, B. Bordass, Flexibility and Adaptability. In designing better buildings, Macmillan, S., Spon Press. 145-156, (2004)
9. A. Friedman, The adaptable house: Designing homes for change. McGraw-Hill. New York (2002)
10. S. Brand, How Buildings Learn: What happens after they're Built, Penguin, New York, (1994)
11. K., Beadle, A. Gibb, S. Austin, A. Fuster, P. Madden, Adaptable futures: setting the agenda in Proceedings of the 1 st I3CON International Conference, Loughborough University, Loughborough, UK, 14-16 May. (2008),
12. P. Crowther, Developing an inclusive model for design for deconstruction, A.R. CHINI, ed. In: Deconstruction and Materials Reuse: Technology, Economic, and Policy, 6th April 2001, CIB, International Council for Research and Innovation in Building Construction, Wellington, New Zealand, (2001)
13. C. Langston, F.K.W., Wong, C.M. Eddie, And H.L., Shen, Strategic assessment of building adaptive reuse opportunities in Hong Kong. Building and Environment, **43(10),** 1709-1718, (2008)
14. C. Langston, On archetypes and building adaptive reuse, Proceedings of 17th Pacific Rim Real Estate Society (PRRES) Conference, Gold Coast, Australia. 16th –19th (2011)
15. H. Estaji, A Review of Flexibility and Adaptability in Housing Design Internat. Jour. of Contemp. Arch. "The New ARCH" Vol. 4, No. 2, (2017)
16. R. Milwicz, P. Nowotarski, Influence of multiphase flexible timber frame house construction on housing affordability, Proc. Eng. **122**,158 – 165, (2015)
17. N.J. Habraken, Design For Flexibility. Building Research & Information, 36 (3), 290-296. (2008)
18. T. Schneider, J. Till, Flexible Housing. Oxford, United Kingdom: Architectural Press, (2007)
19. A. Huber, New Approaches to Housing for the Second Half of Life, Birkhäuser, Berlin, (2008)
20. Ch. Schittich, In Detail: Housing for People of All Ages, Birkhäuser, Berlin, (2007)
21. K. Minami, The efforts to develop longer life housing with adaptability in Japan Energy Procedia (1996)
22. P.A. Jensen, E. Masles, J. Brinkø, Ch. Thuesen, 10 questions concerning sustainable building renovation, Building and Environment **143**,130–137, (2018)

Quality Management Systems as a key element for company strategy selection - case study

Piotr Nowotarski [1,*], *Jerzy Pasławski*[1] , and *Aleksandra Kadler*[1]

[1]Poznan University of Technology, Institute of Structural Engineering pl. M. Skłodowskiej-Curie 5, 60-965 Poznan, Poland

Abstract. Term quality can be defined as the implementation of specific requirements, having its source in the expectations and needs of customers, users of products or services and their buyers. In constructions sector quality is crucial at each and every phase of the process beginning with design ending with the use of the building. There exist many different methods and tools related to quality, Lean Management, Agility, which can enhance construction process. One of such methods is SWOT analysis that can be helpful in determining the current condition of the company performing it. Article presents brief description of the quality concept and introduction to the SWOT method. It is illustrated with the example of analysis performed in company from the construction sector related to design and production of wooden trusses showing which strategy would be most appropriate for the company to execute taking into account its current condition.

1 Introduction

The first concepts of quality assurance in enterprises, mainly in the field of production management appeared at the beginning of XX century. Their rise was caused by the growing industrialization, increasing production capacity of the factories operating at that time and the growing complexity of the manufactured products. The following years brought new techniques and tools as well as the beginnings of standardization in the field of quality systems. Documents were developed - standards - in which, among others, instructions on how to proceed not only to detect but also to prevent errors [1].

In one of the standards - ISO 9000 definition of quality can be found according to which quality is a "degree to which a set of inherent characteristics fulfils requirement" [2].

It can therefore be said that quality is directly connected with meeting requirements, expectations and needs of customers, users of products or services and their buyers. The task of meeting these requirements belongs to suppliers and producers. Using a number of tools and methods, they can effectively perform their work, while having full control over the degree of its correctness - which is understood as being in accordance with the client's expectations. Data collection and analysis enables the possibility to react in case of errors and problems as well as eliminate them at source. An indispensable element of quality

* Corresponding author: piotr.nowotarski@put.poznan.pl

assurance is striving for its continuous improvement. For this purpose the concept of quality management has just been created.

The activities related to quality management are mainly directed at managers, because as participants in the production or service provider, have the greatest impact on the efficiency of the entire process and people directly performing its job. The correct use of their potential, strengths and talents has a significant impact on the quality of the final product. It is especially important in the construction sector processes [3].

Among the treatments falling within the scope of quality management, we can distinguish [4]:

- creating a pro-quality culture;
- planning the resources needed to achieve quality objectives;
- setting the company's quality goals and monitoring the level of their fulfilment;
- defining the responsibilities and responsibilities (tasks) of employees and organizational units for achieving quality objectives;
- creating a system of interpersonal communication;
- motivating employees to get involved in achieving quality goals;
- documenting the requirements and the level of their fulfilment;
- keeping an account and analysing quality costs;
- creating a platform for communication between the enterprise and the environment in matters related to quality problems.

All of the above mentioned aspects are important from the end client point of view and company who owns the process. The thing which bind them together is company management strategy in which client's point of view, as well as current company market position taken into account. According to authors tool which can help in determining proper strategy could be SWOT analysis which is technique used for analyse manage strategic process [5].

Crucial aspect of performed analysis is company's approach to the quality. To one of the 3 fathers of quality [6]: Crosby, quality means "conformance to requirements" [7]. Quality must be defined in measurable and clearly stated terms to help the organization take action based on tangible targets, rather than on bunch, experience, or opinions. Deming, does not define quality in a single phrase. He asserts that the quality of any product or service can only he defined by the customer. Quality is a relative term that will change in meaning depending on the customer's needs [8]. Juran defines quality as "fitness for use.'* He stresses a balance between product features and products free from deficiencies. As used by Juran, the word "product" refers to the output of any process, and that includes goods as well as services [9].

Nowadays researchers point the importance of quality, especially for the construction sector where it (quality) stands together with the cost, and time [10].

The aim of this article is to present SWOT analysis method, which helps in choosing the most appropriate strategy for development of the organization based on the Quality Management practises in the construction industry sector. What is more authors focus on pointing out crucial further recommendations for company to develop and gain success on the global construction market. Article is summed up with the conclusions presenting the most important observations taken during study.

2 SWOT Analysis

The name of this method is derived from the acronym of words: strengths, weaknesses, opportunities, threats [11]. It belongs to one of the longest-used methods of examining the company's internal situation and its position in relation to the competitive environment. The SWOT analysis is a comprehensive method, and thanks to the simplicity of the study, it is also very popular, applicable in various sectors and at many levels of decision making in the company [12]. It originates from the business management discipline and has been widely applied to a broad array of disciplines [13].

Basic assumptions of the SWOT analysis were developed in the 1950s and 1960s by a group of business consultants and scientists working at Harvard Business School. They are then described as "Business Policy: Text and Cases". [14]

Internal analysis is used to designate and assess material and non-material factors that can support the development of an organization or harm its operation, both current and future. The SWOT method is a way of strategic analysis, an algorithm for multilateral assessment of internal and external factors that affect the development potential and the current condition of an organization, company or program [15].

The starting point for the SWOT analysis is the proper recognition of the following factors [16,17]:

- Opportunities - positive external factors - identification of opportunities consists in finding strategic factors in the environment surrounding the organization, which decide about maintaining its position or further development.
- Threats - external negative factors - it is necessary to clarify the major factors occurring in the environment that make it difficult to maintain the position or development of the organization.
- Strengths - positive internal factors - in order to correctly identify strengths one should have knowledge about the analysed organization. The deeper the knowledge of the subject of the analysis, the more effective the result will be. The strengths depend on the type of organization being analysed, in particular its qualifications and resources.
- Weaknesses - internal negative factors - the identification of weaknesses has a similar course as the analysis of strong websites and is the result of knowledge of the subject being studied. In most cases, weaknesses arise from the various shortages of qualifications or resources of the organization.

The SWOT analysis consists of examining four situations answering the following questions [11]:

1. Maxi - maxi (SO)
"Can a specific strength allow me to take advantage of a given opportunity?"
It consists in using the strengths of the organization, e.g. a strong reputation on the market to strengthen external opportunities, e.g. to squeeze new market areas.

2. Maxi - mini (ST)
"Can a specific strength limit the given threat?"
It is based on limiting the adverse impact of the environment using the company's dominant strengths, such as: the ability to innovate on a current basis, a high level of quality, efficient improvement of processes related to customer service;

3. Mini - maxi (WO)
"Does the specific weakness limit the possibility of using a given opportunity?"
It concerns the impact that internal weaknesses (e.g. problems in relations with the client) have on the potential to strengthen opportunities, e.g. building strong competition in a specific environment.

4. Mini - mini (WT)
"Does the specific weakness increase the threat?"

The company operates surrounded by threats that it cannot overcome due to the lack of equivalent strengths. An example of such a situation is, for example, the emergence of so-called negative quality spiral.

Described method in the construction industry can be powerful tool for the strategic planning of the company and with some fuzzy set theory modification can be even in better assisting the formulation of proper strategies [18]. Authors are of the opinion that the most important feature of SWOT analysis is it simplicity, which allows for use of it at different level in the process and in the company. It is especially important in the construction sector, where majority of people involved in actual execution of the process are workers performing their work, who, when comes to the improvements, should understand the idea of proposed methods and solutions.

Case study

Authors decided to perform analysis on the company which was established in 1991 in Poland. From the very beginning, it was related to the wood industry. In the first period, the company specialized in lumber trade, and currently operates as a certified producer of wooden roofs, ceilings and frame houses.

3.1 Process Description

The process chosen for the analysis is related to designing and execution of prefabricated truss structure and consists of several stages. It starts from receiving an inquiry from the customer. The merchant (contact with the client) in consultation with the designer preliminarily assesses whether the given project is feasible in the form of barbed wire plates, used in the company and whether the deadline set by the client is possible to meet. If so, the client is asked to complete the necessary project documentation if the query was incomplete, which includes: the last floor plan, all sections through the building, roof projection and the location of the object. The façades are also helpful for illustrative purposes.

On the basis of the documents sent, the designer sets the architectural layers, and hence the load on the truss. Then model the walls in the program, give the spacing of girders, complements the data on load coverage and climatic conditions. The next step is to model the truss itself. On the basis of the preliminary design project prepared by the constructor, an offer is prepared, which is then presented to the client.

When the customer accepts the offer, the details of the contract are determined, which is then signed. Prior to the implementation of the truss, an assembly and executive project is prepared. It consists of drawings showing the location of the truss elements and concentrations on the plan, as well as all details and details, along with a technical description enabling easy assembly. The project prepared in this way is sent to the customer in order to confirm the compliance of the design assumptions with the actual state of the construction site, to eliminate potential errors related to changes introduced during the construction, which the customer did not notify the contractor of the truss.

After the client has sent back the acceptance, the list of trusses, barbed plates, connectors, braces and files for the saw, i.e. the so-called workshop documentation, which is sent from the design office to the prefabrication plant. In the production hall, using individual programmable saws, individual elements are cut, which are then connected by means of a hydraulic press to the barbed plates.

The finished girders are loaded onto the transport car. On the assembly site, the girders are unloaded and placed on the walls using a crane. Usually they are mounted directly on the rims and connected with it by means of anchors. Special washers are used between the rim and the wood. The attached trusses are connected by means of concentrations.

After the assembly is completed, the customer is given the so-called as-built documentation (approvals, approvals, declarations, certificates). The customer is obliged to sign the assembly acceptance protocol. The signed protocol ends the implementation process.

3.2 SWOT analysis and results

Based on the FMEA, Ishikawa, 5Whys, the tools of Lean Management and Quality Management, as well as direct conversation with the company workers there were 4 groups of factors defined according in the Figure 1.

Secondly authors confront the internal and external factors found, by examining their mutual influence. For this purpose, a zero-one system were used, in which the occurrence of interaction between the factors marked with the number 1, but no dependence - the number 0.

SWOT Factors					
Sym.	Weight	Internal Factors	Sym.	Weight	External Factors
	1,00	**STRENGHT [S]**		1,00	**OPPORTUNITIES [O]**
S1	0,07	a high level of technology corresponding to European standards	O1	0,12	company development and production increase
S2	0,10	price competitiveness	O2	0,11	the opportunity to compete in the industry market
S3	0,07	highly qualified team of designers	O3	0,09	increase in the number of regular customers
S4	0,07	comprehensive offer (design, production, assembly)	O4	0,10	increase in the popularity of the type of construction produced
S5	0,09	large share in the regional market	O5	0,07	technology development
S6	0,05	clear division of duties in individual departments	O6	0,09	long-term cooperation with a design office
S7	0,08	large trust and frequent company instructions among clients	O7	0,11	establishment of a branch in another area of the country
S8	0,10	high quality of trusses	O8	0,09	low reliability of the competition's work
S9	0,09	timely execution of orders	O9	0,11	enlargement of the Staff amount
S10	0,10	many years of experience on the market			
S11	0,08	use of the highest quality raw materials			
S12	0,10	possession of international product quality certificates			
	1,00	**WEAKNESSES [W]**		1,00	**THREATS [T]**
W1	0,11	long waiting times for assembly	T1	0,14	high production costs
W2	0,13	small attachment to the company of production and assembly workers	T2	0,13	the appearance of strong competition
W3	0,11	too few employees	T3	0,09	change in trends (decrease in popularity of technology)
W4	0,09	long waiting time for offers in the season	T4	0,08	increase in unemployment (decrease in the number of investments built)
W5	0,12	narrow range of specialization	T5	0,12	seasonality of production growth
W6	0,14	no quality management system in the company	T6	0,11	a small number of specialized designers
W7	0,14	lack of quality control of products	T7	0,10	loss of trust among customers
W8	0,08	frequent changes to the work schedule	T8	0,11	change of the company owner
W9	0,08	non-compliance with health and safety rules	T9	0,12	closeness to competition

Fig. 1. Identified SWOT factors

Results of this analysis are presented in Figure 2. The list showed that by far the largest number of interactions (170) and the highest sum of products (15,90) has the maxi-maxi strategy. This means that the company is dominated by strengths, the importance of which is important for the company's development. In combination with the large opportunities in its environment, the company may apply an aggressive strategy of further activity, based on strong expansion and diversified development.

Nearly identical results ended with the analysis of the maxi-mini (no. of interactions: 120 and sum of products: 11,82) and mini-mini (no. of interactions: 116 and sum of products:11,28) situation. This indicates a real possibility for the company to limit the impact of external threats by using its strengths and the importance of weaknesses in their enhancement.

	OPPORTUNITIES	THREATS
STRENGHTS	**Aggressive strategy (maxi-maxi)** Number of interactions 170 The sum of products 15.90	**Conservative strategy (maxi-mini)** Number of interactions 120 The sum of products 11.82
WEAKNESSES	**Competitive strategy (mini-maxi)** Number of interactions 92 The sum of products 9.86	**Defensive strategy (mini-mini)** Number of interactions 116 The sum of products 11.28

Fig. 2. Results and strategy for analysed company

The lowest values of the number of interactions (92) and the sum of products (9,86) were achieved by the mini-maxi strategy, which should be interpreted as a small dependence between the weaknesses of the company and its ability to use existing opportunities.

It has to be noted that indicated in the analysis weakness: lack of the quality management systems pays the important role in the current company situation since more and more focus is nowadays paid to the QMS being implemented in the companies – also in construction sector. Having implemented QMS can have positive impact not only on the organization itself, but also on the company-client and company –supplier direction which directly influence the position on the market of analysed company.

4 Conclusions

Analysis of the presented process revealed both weak and strong sides of the company. By pointing out problems and examining their importance several main problems that threaten the proper execution of the roof truss were examined.

Bearing in mind the great potential of the company, which has been demonstrated by the SWOT analysis, it is proposed to use an aggressive strategy as part of planning future activities. The company has the opportunity to conquer new markets, not only Polish but also foreign ones, and to increase production if it uses its strengths wisely.

In details, SWOT analysis conducted for the company taken into account showed that:

a) the enterprise has a chance to develop not only on the Polish market, but also to expand its activity to foreign markets. Bearing in mind the cost of producing a truss in Poland and the quality offered by the company, there is a great chance of expanding into European markets, even in spite of the popularity and availability of precast concrete truss technology in other European countries.

b) Introduction ISO 9001 standard can definitely help in proper organization of the processes in the company as well as can influence the position on the market.

c) the company is struggling with the lack of involvement among manual workers, which manifests itself lack of interest in training, frequent staff changes or absenteeism from work.

d) despite the lack of knowledge or application of quality management systems, the company has many strengths. In the opinion of clients, the company is characterized by high quality of manufactured beams, which translates into high trust and frequent recommendations of the company's services.

The advantages mentioned above based on the SWOT analysis, according to which the aggressive strategy of conquering new markets and developing presented directions is the best strategy for further action for the company.

Authors are planning further investigations in presented research area with special attention to the processes improvement synergy with the use of Lean, Quality 4.0, and Agile Management principles in SME companies, which can lead to the significant continues improvement of construction processes.

Acknowledgements

The publication was created with the support of statutory funds of Institute of Structural Engineering at Poznan University of Technology

References

1. A. Mazur, A., H. Gołaś, H. *Zasady, metody i techniki wykorzystywane w zarządzaniu jakością*. (Poznań University of Technology, Poznań 2010)
2. ISO standard 9000:2015- Quality management system – Fundamentals and vocabulary
3. Z. Ma, S. Cai, N. Mao, Q. Yang, J. Feng, P. Wang, Construction quality management based on a collaborative system using BIM and indoor positioning. Automation in Construction, 92, 35-45 (2018)
4. A. Hamrol, *Zarządzanie i inżynieria jakości*. (Wydawnictwo naukowe PWN , Kraków, 2017)
5. S. Khaba, C. Bhar, C. Quantifying SWOT analysis for the Indian coal mining industry using Fuzzy DEMATEL. Benchmarking: An International Journal, 24(4), 882-902 (2017)
6. J.G. Suarez, *Three Experts on Quality Management: Philip B. Crosby, W. Edwards Deming, Joseph M. Juran* (Total Quality Leadership Office Arlington V, 1992)
7. P.B. Crosby, Q. I. Free, *The art of making quality certain*. (NY: New American Library,1979)
8. W. E. Deming, *Out of the Crisis*. MIT press (2000)
9. J. Juran, G. M. Gryna, *Juran's Quality Control Handbook*. (NY: McGraw-Hill 1988)
10. M. Gajzler, K. Zima, Evaluation of planned construction projects using fuzzy logic. International Journal of Civil Engineering, 15(4), 641-652 (2017)
11. M Szczęsna, D. Klimecka-Tatar, D.. Analiza SWOT w zakładzie produkującym cement. Archiwum Wiedzy Inżynierskiej, 2, 17-19, (2017)
12. K. Cholewicka-Goździk, Opłacalność jakości. Problemy Jakości, (7), 7-12 (2009)

13. H. Yuan, A SWOT analysis of successful construction waste management. Journal of Cleaner Production, 39, 1-8 (2013)
14. K. Woźniak, M. Sołtysik, *Analiza SWOT* online: https://mfiles.pl/pl/index.php/Analiza_SWOT (accessed October 11th, 2018)
15. Ł. Szałata, J. Zwoździak,. Analiza SWOT jako podstawowe narzędzie w zarządzaniu środowiskiem. Rocznik Ochrona Środowiska (Annual Set The Environment Protection), 13, 1105-1114 (2011)
16. J. Nazarko, J. Ejdys, K. Halicka, A. Magruk, Ł. Nazarko, A. Skorek, Application of Enhanced SWOT Analysis in the Future-oriented Public Management of Technology. Procedia Engineering, 182, 482-490 (2017)
17. B. Igliński, A. Iglińska, G. Koziński, M. Skrzatek, and R. Buczkowski, "Wind energy in Poland – History, current state, surveys, Renewable Energy Sources Act, SWOT analysis", Renewable and Sustainable Energy Reviews 64, 19–33 (2016)
18. M Gunduz, M.T. Birgonul, M. Ozdemir, Development of a safety performance index assessment tool by using a fuzzy structural equation model for construction sites. Automation in Construction, 85, 124-134 (2018)

Characteristic of materials for the 3D printed building constructions by additive printing

Katarzyna Pacewicz[1,*], *Anna* Sobotka[2], and *Łukasz* Gołek[1]

[1,2,3]AGH University of Science and Technology, al. Mickiewicza 30, 30-059 Cracow, Poland

Abstract. Three dimensional printing is a promising new technology to erect construction objects. Around the world in every moment a new prototypes constructions are made by using this method. Three dimensional printing is taken into account as technology which can be used to print constructions in automated way on the Moon or Mars. The raw materials, which can be used with three dimensional printing have to fulfil basic requirements for those which are used in construction. That means that components of printing mortars are made from ingredients easily accessible in area nearby construction site and can be reusable. The cost of printing building objects due to that requirements is comparable to costs of traditional building, which are currently available. However additive techniques of printing needs a dedicated mortars for printer supplying. Characteristic for such mortars is: setting time, compressive strength, followability in the printing system, shape stability of every printed layer, controlling the hydration rate to ensure bonding with the subsequent layer, reusable capabilities, easily accessible raw materials, cost of such mixtures shouldn't be too high in order to keep 3D printing competitive for traditional ways of building, mortar components should be recyclable and printing process should not influence negatively on an environment and people. All properties of printing mortars are determined by the device for additive application method. In this paper review of available materials used for three dimensional printing technology at construction site is presented. Presented materials were analysed in terms of requirements for building materials technology. Due to the lack of detailed information's in available literature, regarding to the properties of raw materials, the results of this analysis may be used in the designing of new concrete mixtures for the use in three-dimensional printing technology for construction.

1 Introduction

Three dimension printing is a process of making three-dimensional, physical objects based on computer model. At the beginning this technology was used to make device's prototypes and machine's elements. The development of three-dimensional printing technology accelerate the process of designer's vision became true.

Commercialized and widely admitted the 3-dimensional printing technology is stereolithography, which was patented in 1986 by Charls Hull. His method uses local

* Corresponding author: kwronska@agh.edu.pl

hardening photopolymer resins by a strong laser light beam. In 1989 Carl Deckard patented Selected Laser Sintering (SLS), which also basis on hardening material by laser beam. This solution is very similar to stereolithography, however powder supplements s needs to be used. The most popular 3-dimensional printing technology is Fused Deposition Modeling (FDM), which was patented in 1992. In FDM method the thermoplastic material is extruded by heated by printer's nozzle. The nozzle and temperature controls the melted material flow. Object is created layer by layer. Firstly the object contour is drawn, than the created area needs to be filled up. In order to solve issue with elements which needs supports the additional printhead is used. One printhead provides building material, while the second provides material for support, which can be water-soluble or separated from printed element. FDM uses thermoplastic materials, with high mechanical strength, thermal resistance and chemical durability. 3 Dimensional printing (3DP) puts thin layer of powdered composite, which is blended with the sprayed binder, from the printer's head moved in horizontal plane. When the layer is finished, roller puts new layer of powder material on lowered machine's table. In this technology the additional supports are not required. It allows to build complex structures, if the object thickness and its strength allows to safely remove it from working chamber. Objects made with 3DP method are fragile and damage sensitive. In order to increase their mechanical strength, appropriate treatment is performed and then hardener and UV protection agents should be applied. The 3DP is currently available in three different variants, where as supplements the ceramic, plastic and powdered metal materials are used [12].

Currently a lot of constructions around the world are built with 3-dimensional printing technology and China is leading this industry branch, however there is a lack of information about used formulas and mixtures, which could be used for printing technology. Therefore, the authors, based on the analysis of the available literature and the analysis of the concrete mixtures, attempted to combine the required characteristics of materials and mortars that provide the ability to use it as supply materials for printers and meet the key requirements such as the compressive and bending strength. This article contains the description of chosen implementations and materials used for it, such as raw materials and mortar mixtures. The influence of mortar components on the properties of fresh and hardened mortar were analysed. Results can be used to design new concrete recipes for construction printing.

2. Three dimensional printing process in construction

The print process of the ready element consists of seven stages. Stage one is the graphical 3D model computer designing (mostly CAD). Next step is conversion of prepared model into format, which is understandable by 3D printer. Usually this is STL format. Before printing, the printer is adapted to work and the drawing file is adapted to set the print scale and orientation of printed items. Following stage is 3D object printing. Material is laid layer by layer with specified thickness, according to 3D model. When the printing process is finished, printer needs to be clean up. Last stage is object's finishing.

Automation of building processes can result in a reduction of the work force, which increases safety at the construction site, reduces construction time and the production costs. In addition, the development of building information modelling and the use of digital information streamlines the automation and production process in construction. Currently, there are several projects related to creation of constructions in the additive technology, namely: Apis-Cor, Contour Crafting, D-Shape, Concrete Printing.

Russian's company Apis-cor [1] has printer capable to work at construction site. Printer's size is 4m/1.6m/1.5m and weighs 2 tons (figure 1). The printer is placed on one base (foot). A special support system does not require the terrain preparation. Device work like construction crane, that allows printer to rotate and print object's inside and outside. Printer's work space is 132m^2. Mixture for printing is based on cement.

Final hardened material from such recipe indicate the properties like the standard C16/20 concrete. In this project the geopolymers and fiber concrete are also used. The foundation is firstly created by drawing the contour, next it's reinforced and filled up with concrete. After the foundation and insulation works, the contours of the walls are printed and reinforced with the zigzag shape inside the wall.

Cooperation with Geobeton company brings a new recipes for concrete to use in construction industry. Geocement^{TM} consists of two components: powder geopolymer cement and the liquid reagent – geosilicate. Additionally, the company offers dry geopolymer concrete, which is delivered at construction side in custom amounts for concrete production. Geopolymer structure and high level of mesoporosity brings the air flow through material, so it creates insulation, frost and fire resistance. Some other properties are very adjustable, like thixotropy or slump. The setting time in very wide range, up to 180 minutes. Geopolymer concretes indicate high compressive strength, over 100 MPa and chemical resistance on alkalis and acids. On the basis of a geopolymer binder, various compositions of geo-concrete were developed and tested. The tests were carried out in Irkutsk, Russia. The industrial by products were used in the developed recipe [1].

Fig. 1. Printing process presented in February 2017 in Stupino city with Apis Cor construction 3D printer [1]

The Contour Crafting project was developed by the team of professor Behrokh Kloshnevisfrom the University of Southern California. It is a device mounted to the crane frame. The robot's additional arm installs appropriate channels to supply the facility with media and horizontal structural elements, which are difficult to print. This 3D printing technology involves the additive application of cement mortar with a trowel-tipped head. The trowel usage allows to smooth the unevenness of the side surfaces, which are result of layers addition. Currently, the head is capable of additive construction of a full-sized structural wall, with the minimal consumption of material [6,12].

The D-Shape printer can print monolithic structures (figure 2). It uses a powder deposition process, which is selectively cured by means of a binder. In this method, the printing process consists in arranging the appropriate thickness of the building material, compaction of the layer and application of the binder by the nozzle mounted on the ramp. After the finished printing process, the printed object is extracted from the loose powder bed. The dry material, which was not treated by the binder is used as a support for the constructed objects, additionally it can be used in the further printing process [5,6].

Cementitious Micro-Concrete printing process uses materials with various granulation, from 0.2mm to 5mm. Material binding can be carried out with water solutions or cement binder. The inert raw material may be: crushed stone, sand, gravel, crushed clay aggregates and ceramics. Also recycled materials can be used, such as: glass, wood chips, shredded plastics, demolition materials, rubber, gypsum or a mixture of the above mentioned batches, reinforced by fibers: natural, glass, basalt, polymer or steel. As a liquid binders and reagents for inert batches are used: suspensions based on Portland cement, Sorel cement or geopolymers, polymeric primers (urethane, silossane, vinyl with the exposition of fenolic and PBB, PPC binder) all with $v \leq 3°E$ [2].

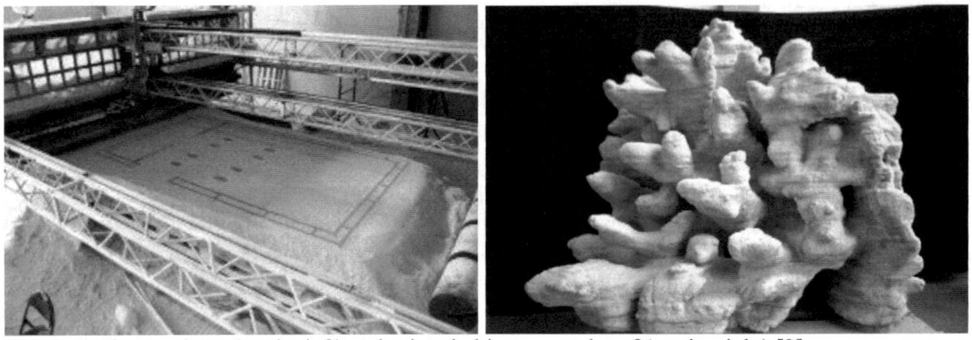

Fig. 2. D-Shape printer (on the left) and printed object – coral reef (on the right) [2]

Examinations on "Concrete Printing" project are carried out by Loughborough University [4,5,6,7]. Printing device prototype is made from steel frame with dimensions 5.4m length, 4.4m wide, 5.4m height and printing head is mounted to moveable horizontal bar (figure 3). Printing process can be described in three steps: data preparation concrete mixture preparation and printing. Prepared mixture is placed in the pump on the external platform, in order to feed it through the printing nozzle with diameter 9mm. The proposed solution provides small buffer tank, placed on the top of printer.

This project uses two kinds of materials for printing. The cement-based material is used to build the elements, while the gypsum-based material serves as support material for the printed element. Gypsum material was used in this project due to easy material removal, low mechanical strength and 100% recycling. Retard admixtures are added as a mixture components to provide required open time and to obtain a constant workability of the material during production.

Fig. 3. Concrete printing system (on the left) and printed bench (on the right) [4]

The research described in paper [4] uses sand with a maximum grain size of 2mm. As a binding components the cement CEM I 52,5, fly ash and non-compressed silica dust were used. The dry components were added to mixing water together with polycarboxylate-based superplasticizer to increase the workability and material strength. In order to obtain the proper open time, the retarder based on amino-tris (methylene phosphonic acid), citric acid and formaldehyde was added. Studies were also carried out using an accelerator based on sulfones, aluminum salts and diethanolamine. The concrete also contained polypropylene microfibers with a length of 12mm and a diameter of 0.18mm, which were designed to reduce shrinkage and deformation in the plastic state.

3. Characteristic of mixture's components for additive printing

Presented examples of 3D printing projects of buildings and used materials are not described in details, but allows to analyse the required features of a mortar mixtures that can be used in 3D printers for printing of building objects. For 3D printing the research teams, mentioned in the article, mostly used cement-based materials. Of course the material used for 3D printing has to be compatible with 3D printer. In the composition of modern concrete, additives and admixtures are present along with aggregates, cement and water, which are added to improve selected properties of mortars.

Additives can significantly modify properties of fresh or hardened mortar. While designing mixtures of mortars useful for 3D printing, the care should be taken that the mortar meets the performance assumptions of fresh and hardened mortars. When designing the composition of mortar, to the necessity of extrusion should be taken into account, what is influencing on workability and open time. The key properties of hardened concrete are compressive and bending strength.

Additives used to produce concrete include: ground granulated blast-furnance slag, fly ash, silica fume, and rock powders: limestone and quartz powders, which are treated as chemically passive components.

Fly ash – its wide application is determined by its high fineness, chemical and phase composition as well as pozzolanic activity. Fly ash improves workability of mortar mixtures and significantly slows down mortars hardenıng process. Concrete shrinkage decreases in proportion to amount of cement replaced by fly ash. Fly ashes with high content of unburned coal lowers the effectivity of chemical admixtures, especially aeration, plasticizers of superplasticizers. Silicoeous fly ashes improves mixtures workability, that improves its pumping process.

Ground granulated blast furnace slag is obtained by rapid cooling of liquid slag, which is formed as a by-product in the production of a pig iron in a blast furnace. Then it is milled, with the agents facilitating its grinding which should not exceed 1.0%. The organic components should not exceed 0.2% of the slag mass. Slag is more active and more effectively seals concrete in relation to fly ash.

Silica fume is a by-product of the production of metallic silicon or its alloys. Silica dust seals the microstructure of concrete by filling the space between cement grains. The use of silica dust in an amount of 2-3% (as an admixture) of cement mass, has a beneficial effect on improving workability of the mixture along with good viscosity. Silica dust forms a sol with water, what prevent water outflowing of the mixture while compacting. Rock powders improve the workability of mortars. The most beneficial effects are obtained by using powders obtained from low-crystalline limestone rocks. Bentonite, as a clay rock, improves the tixotropic properties of mortar mixtures. This material contains mainly of aluminosilicates, which influenced by water swell and increase their volume several times.

Synthetic resins show a significant adhesion with aggregate and hardened mortar. A properly selected composition of resin improves workability of the mortar as well as mechanical strength and physical properties [3,8].

The role of admixtures is to improve one or several properties of a mortars or hardened concrete. They are added in amount up to 5% of the cement mass, during mixing. While designing mortars for 3D printing, it is important to focus on effects of modifications caused by admixtures, which can improve one and worsen the another. Superplasticizers are based on the synthetic resins that give a negative charge to the cement grains, which facilitates the formation of a water layer between the cement grains. Due to the interaction of electrostatic forces between the molecules, the workability of mixture is improved. Mixture internal friction and surface tension of the water are reduced. The process of wetting the aggregate and binder is easier, and the mortar mixtures becomes more plastic.

The selection of the right superplasticizer depends on the cement properties and also on additives and on aggregates (to some extent). Difficulty in selection results from the differences between the binder and the superplasticizer properties.

Admixtures that accelerate hardening do not cause violent binding reaction, only shorten it, which accelerates the hardening period. These admixtures are designed to increase the dynamics of mortar strength increase over time. Process of initial strength increase is accelerated, with no negative effects on the final mortar strength. While printing successive layers of the object, the initial mortar strength is very important to prevent delamination and printed structure disintegration.

Retarding admixtures are used because during the process of printing the transport of the mixture takes a long time and it is needed to obtain proper bonding between particular mortar layers. Retarding of the bond causes the transported mixture to retain its liquidity for longer time, which has a positive effect on the pumping and printing processes. Concrete with a retardant admixture is indicate lower initial strength and usually slower hardening.

The viscosity modifying admixtures are designed to increase the viscosity of mortars, without causing significant changes in consistency. They are usually used to eliminate segregation, especially in self-compacting mixtures. Consideration should be given to the technology of feeding the mixture to the 3D printer, because while pumping the mixture, increasing the viscosity requires a higher pumping pressure [3,8]

4. Additive printing mixes conditions and specific features

Based on previous research and experiences of 3D printing technology in construction, especially for building structures, it is possible to determine the principles the composition of mortar mixtures and its required properties (Fig. 4). It should be taken into account many different implementations, conditions and variants of buildings, in particular the specific conditions of the construction site. While designing a mortar mixture intended for use is 3D printing technology, it should be noted that this mixture meets the accepted criteria and design assumptions.

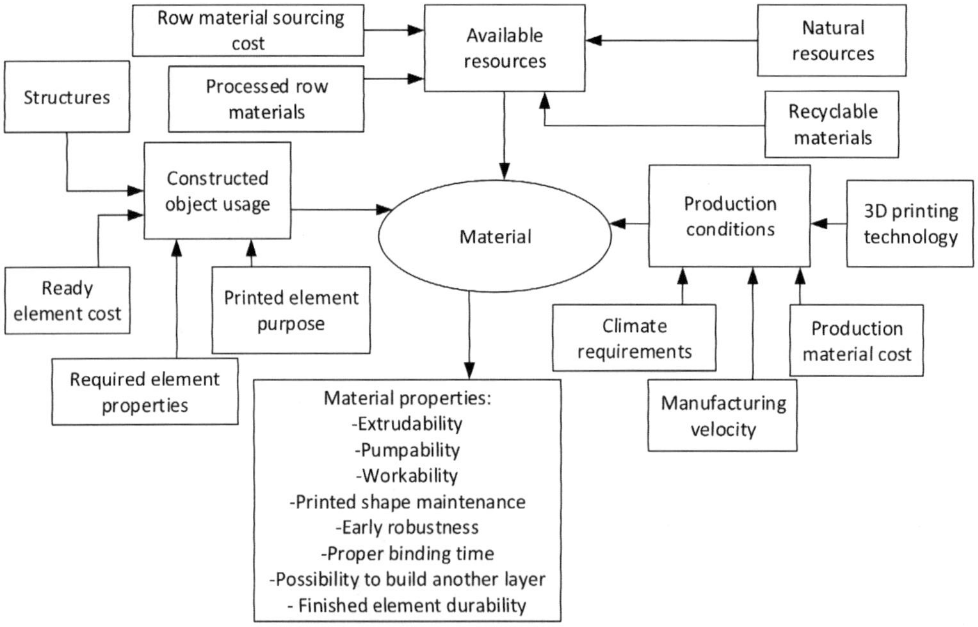

Fig. 4. Building material composition for 3D printing and properties requirements

Material, designed and used in 3D printing technology for construction, depends on three interpenetrating factors: raw materials, methods of application and methods of production. While designing such a material, it is also necessary to pay attention to the properties that the manufactured object has to fulfil and that it is possible to print by the given printer. Designer should pay attention to the locally available raw materials, near the construction site, so that the cost of their transport would make economic sense. The composition of the mixture is also influenced by climatic conditions at the place of object's production, as it affect its composition and maturing conditions. The time after which the given properties will be achieved also affects the printing process, it also depends on the method of material delivery and the time of applying subsequent layers of building material. The composition of the material also depends on the cost of producing a given object.

As stated above, while designing and performing 3D printing from a mortar mixtures, it should pay attention to its properties that affect the possibility and quality of the printed object. Particular attention should be paid to such parameters as: flowability of the mixture, extrudability, stability of the printed shape, bonding time, strength and shrinkage of concrete.

The flowability of the mixture allows the pumping the mixture along the feeding and mixing systems. Regulation of the mixture rheology and its flow in the fresh state is ensured due to the wide particle size distribution. The use of different granulation of additives contributes to a denser packing of the mix grains, which results in better fluidity. It should be noted that too much dusty fraction (increase of intermolecular friction) can increase the viscosity of the mixture, which will negatively affect its fluidity.

Extrudability is the ability of the mixture to be transported over the small tubes and nozzles in the print head. The workability and consistency of the mixture have a fundamental influence on the extrudability. In order to obtain a good extrudability of the mixture, concrete should be designed in accordance to the principles of designing self-compacting concrete and spraying concrete.

The ability to preserve printed shape is another important parameter that should be met by the pumped mixture used for 3D printing. The printed layers of material should keep their shape formed by nozzles under the pressure of subsequent layers applied by the printer. The decisive parameter is the early strength of the hardened mixture. The shape stability is ensured by increased amount of fine aggregate and sand. Viscosity modifying admixtures can reduce the overflowing of the mixture and improve the stability of the printed structure.

The setting time affects the possibility of mixing during printing. Workability, which affects possibility of extrusion, needs a long "opening time". Otherwise, the material, becomes stiffer and causes the slower flow of the mixture through the feed systems of the printer, or blocks the printer. Additionally, the long open time helps in connecting material between the layers, but it can negatively affect the formation of printed elements. Short bonding time is needed while laying successive layers of material, to keep the shape and load of the higher applied layers. The strength and stiffness of the printed shape is obtained due to the increased density of the matrix obtained by using of finer powders than Portland cement. The strength of the mixture is also influenced by: water-cement ratio, cement strength class, aggregate quality and the presence of admixtures and additives. The contraction of the concrete may start already in the binding period as well as in the initial hardening period. Shrinkage significantly affects the dimensional accuracy and stability of printed objects. Shrinkages are caused by evaporation of water from external layers and changes of the temperature increase caused by heat of hydration. Printed elements have a much larger surface exposed to the influence of ambient conditions in relation to the elements made with the formwork. While designing, it should be noted that the shrinkage increases, among others, with the increase the amount of water in the past, cement fineness and grain size of aggregate.

5. Summary

3D printing technology development in the industry has contributed to the large technological jump. Thanks to various methods of object production, 3D printing enables testing on prototypes, previously difficult to perform with normal production methods.

This paper focuses on 3D printing technology currently used in construction. 3D printing of mortar mixtures is a promising method that can revolutionize the traditional building processes. Safety for building is increased through highly effective automation of the building process, and the demand for workforce is reduced. The costs of creating of new construction objects can be much cheaper compared to traditional building methods. The basic issue in 3D printing of buildings is, except the printer, the material for printing. Many new materials have already been developed, among others: biological origined, steel, wooden and plastics, allowing additive printing of products. However, it seems that for each project designed for implementation in additive printing technology, an individual composition of the printing mixture should be designed.

The designed mixture for printing the object first of all should have certain specific features that enable printing and standard requirements that must pass the particular elements of the building. Due to that whole building could meet the required standards, like strength, fire protection, etc. On the other hand, the raw materials for the production of mixtures should be selected in accordance with the concept of sustainable development, ie saving non-renewable resources, waste materials that can be recycled, including possible minimization of costs. Modelling the mixture properties of, selection of its properties also results from the function that printed construction or the element in the printed object will perform. Also, implementation conditions, including climatic conditions, may affect the required characteristics of the used mixture.

Mixtures for printing should have an advantageous printable properties that can be controlled in the wide range and modify their properties to improve performance during the 3D printing process. Preparing a compatible material for a large-scale 3D printer is a difficult task. There are various factors, which affect on the properties of fresh and hardened mortars. The detailed analysis of the properties and consideration of the mechanisms is definitely recommended due to action of individual additives and admixtures affect on the workability of the printing mixtures. Also due to the capabilities of the printer. Although this technology is in experimental phase, 3D printing from mortar mixtures can revolutionize and increase the potential of the construction sector.

References

1. Apis cor, w:http://apis-cor.com available 15.04.2018.
2. D-Shape, w:https://d-shape.com available 15.04.2018.
3. Z. Jamroży, *Beton i jego technologie*, *PWN*, Warszawa (2015)
4. T.T. Le, S. A. Austin, S. Lim, R. A. Buswell, A. G. F. Gibb, T. Thorpe, *Mix design and fresh properties for high-performance printing concrete*, *Materials and Structures*, **45**, 1221-1232, (2012)

5. T. T. Le, S. A. Austin, S. Lim, R. A. Buswell, R. Law, A. G. F. Gibb, T. Thorpe, *Hardened properties of high-performance printing concrete*, *Cement and Concrete Research*, **42(3)**, 558 – 566, (2012)
6. S. Lim, R. A. Buswell, T. T. Le, S. A. Austin, A. G. F. Gibb, T. Thorpe, *Developments in construction-scale additive manufacturing processes*, *Automation in Construction*, **21(1)**, 262-268, (2012)
7. S. Lim, T. Le, J. Webster, R. Buswell, S. Austin, A. Gibb, T. Thorpe *Fabricating construction components using layered manufacturing technology*, *(GICC'09)*, Loughborough University, pp. 512-520, (2009)
8. G. Ma, L. Wang, *A critical review of preparation design and workability measurement of concrete material for largescale 3D printing*, *Frontiers of Structural and Civil Engineering*, 1-19, (2017)
9. Z. Malaeb, H. Hachem, A. Tourbah, T. Maalouf, N. El Zarwi, F. Hamzeh, *3D Concrete Printing: Machine and Mix Design*, *IJCIET*, **6(6)**, 14-22, (2015)
10. B. Nematollahi, M. Xia, J. Sanjayan, *Current Progress of 3D Concrete Printing Technologies*, *ISARC 2017*, 260-267, (2017)
11. V. N. Nerella, M. Krause, M. Näther, V. Mechtcherine, *Studying printability of fresh concrete for formwork free Concrete on-site 3D Printing technology (CONPrint3D)*, *25th Conference on Rheology of Building Materials*, Regensburg, 236 – 246, (2016)
12. A. Sobotka, K. Wrońska, *3D w budownictwie, Cz.1,2 —Capabilities of use the three-dimensional printing technology in construction industry, Builder,* **11,12**, pp 26-29, (2015)

Concrete pavements management

Łukasz Rudziński[1,2] *Mieczysław Słowik*[3]

[1]Autostrada Wielkopolska II SA, Dziadoszańska 10 Poznań, Poland
[2]Doctoral student, Poznan University of Technology, Piotrowo 5 Poznań, Poland
[3]Poznan University of Technology, Institute of Civil Engineering, Piotrowo 5 Poznań, Poland

Abstract. The article discusses the issue of managing roads with concrete surfaces. Modern methods of surface features identification applied on the exploited motorway section of a length of 100 km (in conversion to rows of concrete slabs it gives around 600 km) are presented. The general principles of the rigid surface management system are described below in order to enable a comprehensive analysis of the obtained data in terms of the possibility of developing concrete slabs degradation models that form the basis for long-term planning of maintenance works.

1. Introduction

Portland cement concrete is becoming increasingly used as a road paving material in many projects on a number of key sections of the major highways of the Polish road network (motorways and trunk roads). This makes appropriate maintenance a key issue, so that the design lifetime of 30-40 years could be achieved with minimum maintenance requirement in the first phase of pavement life [1].

Contemporary pavement distress evaluation methods, including in particular surface conditions identification techniques are not commonly known in Poland and should be popularised in view of the increasing number of roads with concrete pavements and prepared (designed to accommodate different concrete slab geometries with the slab width exceeding the width of the traffic lane) to ensure the required quality of the road network data without compromising the efficiency and maintaining the non-intrusive nature of the pavement condition surveys. An important part of the pavement management process during its service life is appropriate monitoring of the pavement condition consisting in collecting condition

[1] Corresponding author: lukasz.k.rudzinski@doctorate.put.poznan.pl

evaluation input data which are critical for reliable residual life modelling. This knowledge is indispensable for planning the maintenance treatments and renewals, defining priorities and allocation of funds in the long term. The necessary inputs for pavement condition modelling include evaluation of its current condition. Together with the identification of the causes of distress it is the basis for making pavement condition projections. Failure to carry out maintenance treatments at low severity level of distress at an appropriate time can considerably increase the cost of maintenance necessary to repair seriously deteriorated pavement. Figure 1 shows typical concrete pavement curves presenting the periods in which individual treatments and curves extending the life cycle of the pavement are performed through appropriate management of maintenance procedures.
The aim of the presented research work is to develop a solution that allows obtaining the longest life cycle of a concrete surface by analyzing surface images of individual concrete slabs and using them to plan maintenance treatments.

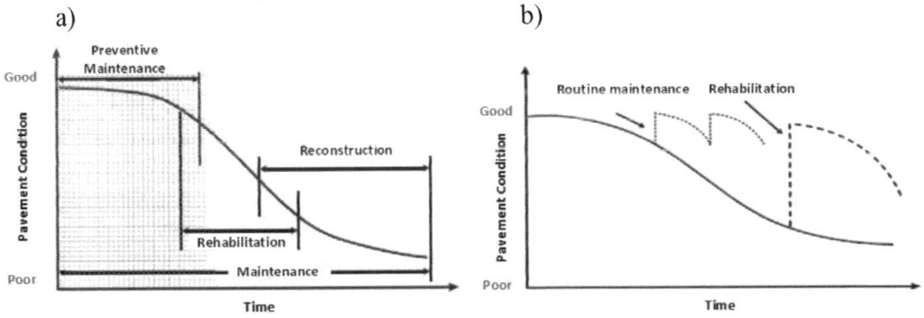

Fig.1. Typical concrete pavement curves presenting the periods in which individual treatments are performed (a) and curves extending the life cycle of the pavement through appropriate management of maintenance procedures (b) [2]

2. Contemporary identification method of the surface conditions of concrete pavements

In road diagnostics we can speak of a successful introduction of equipment or technologies no sooner than after they have been verified under traffic [3]. For this reason, in this presentation of the contemporary surface conditions identification technique its reliability has been verified on a ca. 100 km long section of concrete pavement with dowelled and tied joints (ca. 600 km total length of concrete lanes) consisting of the total number of 121,185 slabs of 4x5m and 3x5m in size.

Bearing in mind that a contemporary distress evaluation method should be based on repeatable measurements and taking into account errors resulting from different readings of the on-board GPS and distance measuring devices (in the range of up to 10 or 20 metres) between different surveys (which are, however, acceptable for most maintenance management procedures) it is necessary to assign each detected distress to a specific slab. To this end a dedicated slab layout map has been prepared including appropriate referencing of the pavement slabs. All the concrete slabs have been identified with x, y co-ordinates and a unique ID ensuring repeatability relating the respective slabs and enabling their prompt and definite identification by the maintenance services assisted with Trimble system (Fig. 2). This ensured repeatability of measurements at the slab level. This approach is necessary to ensure synchronised data of appropriate quality that can be used for further verification of the condition of pavement. It also supports detailed life-cycle analyses at the level of individual slabs rather than over a dozen metres long road segments. Moreover, with this method of identification it is possible to avoid interfering with the equipment installed on the motorway.

Fig. 2. Slab identification kit

As a part of the conducted diagnostics of the concrete pavement, a research methodology was developed, using a special vehicle recording surface conditions of the pavement. The pavement was surveyed with the System for Pavement Distress Evaluation (SPDE) at the disposal of the Polish Road and Bridge Research Institute (IBDiM) (Fig. 3a). SPDE provides digital images of the pavement surface and any distress it contains over up to 3.9 m wide strip with 1 mm x 1 mm image resolution. It was designed to enable surveying road network in any traffic conditions. LED lighting technology makes the technique insensitive to any variation of lighting conditions. Flashes of LED lights have a high magnitude in the recorded spectrum, several times the daylight level. There is also a right-of-way camera installed in the front of the lane.

The images obtained with this special equipment were additionally processed and calibrated to enable geometrical measurements of the identified distresses. This was a necessary step to ensure accurate identification of both surface and sub-surface types of distress.

Image calibration was performed using affine transformations and image filtering and pattern recognition algorithms were used to detect transverse and longitudinal joints (Fig. 3b) [4].

a) b) c)

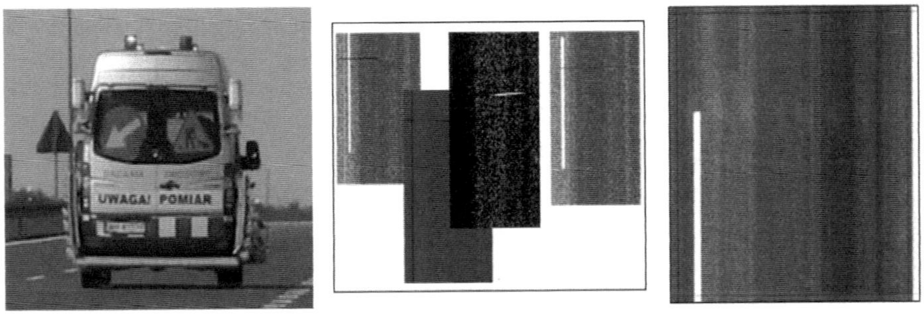

Fig. 3. The process of collecting the data representing the surface conditions of concrete slabs.

Since the concrete slabs in the traffic lanes exceeded the maximum operating width of the instrument it was necessary to synchronise the data from adjacent passes. Then the data were merged to cover the whole concrete lane and to retrieve the respective images. For accurate analysis of the coded distresses, macro images were processed in such a way that a single image showed a single slab with straight vertical longitudinal edges which were used for coding the geometry of each distress. The images have a margin of 15-20 cm which enabled full evaluation of pavement distress, including longitudinal and transverse cracking, corner breaks, edge cracking, spalled joints, defective joint sealing, surface damage and patches.

The example slab (Fig. 3c) is 4 m wide by 5 m long and the image size is 1568 x 2571px. The slab layout map was used in the final step for quick verification of the field data.

3. The use of analysis results in the concrete pavements management system

The starting point for developing the functionality of an efficient concrete pavement distress evaluation system should be reliable, accurate and up-to-date information on the condition of pavement which is the fundamental basis of regular maintenance of road network. These data provide also the input for short and long-term planning of maintenance treatments. They give feedback on the effectiveness of different pavement treatment types and technologies. Moreover, they enable estimating further deterioration of the pavement and creating on this basis pavement deterioration models used in planning of expenditures in the long term.

Therefore, in response to the demand for a system that would provide easy access to the required data, it was necessary to create a consolidated database including display function that would enable easy loading of the digital imaging data (grouped by survey) and, in particular, identification of the surface distresses with layout map reference to be used for subsequent monitoring. The windows presented in Fig. 4 are mutually synchronised such that when a selected location is changed in one window, the contents of the remaining windows is changed accordingly. This enables combining and analysing in the chosen way the displayed information from the different surveys (both historic and current).

The database consolidated in this way supports their further processing, including without limitation measuring the length, width and surface areas of the identified distresses. With these operations, the condition evaluation performed with such distress evaluation system is in compliance with the requirements of SOSN-B [5] and DSN [6] guidelines.

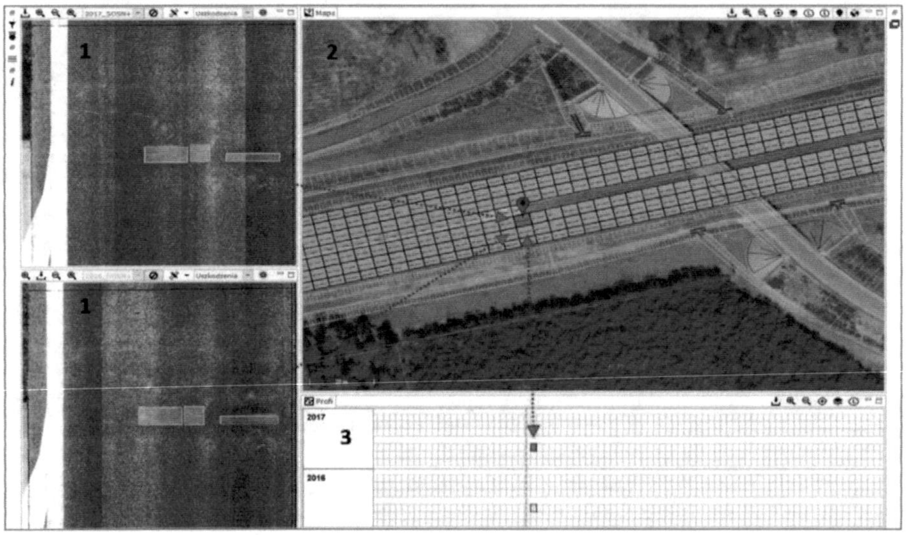

Fig. 4. An exemplary view of the concrete pavement management system window, 1. Macro photos of single concrete slabs in relation to a specific measuring campaign. 2. Geodetic map with marking and location area of all individual concrete slabs. 3. A thematic profile with appropriately defined data

Moreover, replacing the inspector's subjective assessment with objective evaluation of the distress geometry and size data (width, length, area) accompanied with display of values on the unit screen (note that this enables conducting case studies on large populations) makes it a reliable surveying method free of subjectivity related deficiencies typical of visual surveys. This approach enables also analysing the changes in the condition of pavement between surveys and after repairs for analytical and verification purposes (Fig. 5).

Fig. 5. View of the macro images of the same concrete slab for analysing the progress of changes to surface repairs in the subsequent surveys

4. Conclusions

Experimental aspect presented in this paper is a big step towards quality improvement in the concrete pavement management systems. With precise analyses at the individual concrete slab level (taking into account full scope of geometric data) and constant updating of databases with the data obtained through automated pavement condition imaging procedure it is possible not only to obtain "static information" but also, by introducing the time factor, to determine the rate of change of the pavement condition being the necessary input for modelling its residual life. The knowledge acquired in this way will directly influence the timing and type of response of the maintenance teams making it more efficient both in selection of maintenance treatments and in planning the maintenance policy itself.

The distress evaluation system designed in this way in relation to concrete pavements will provide, over a long-term period of observation a unique set of data which can be used for analyses (studying a large population, in this case more than 121 thousand slabs) which could not be done until recently.

What is important is that this concrete slabs management method will support decision making in relation to maintenance treatments and verification of their quality. Moreover, it will provide perfect data for predicting the deterioration models for each type of distress and the applied repair technique.

References

1. Glinicki M.A., Durability of concrete in road pavements. Effect of microstructure, materials design and diagnostics. IBDiM, Warszawa, 2011

2. Concrete Pavement Preservation Guide, Second Edition, September 2014, FHWA Publication.
3. Sudyka J., Mechowski T., Harasim P.: Modern pavement diagnostic techniques in road network maintenance. IV International Conference „Modern Technologies in Highway Engineering", Poznan 2009.
4. Bukowicki M., Opara K., Exact localization damages on concrete slab pavements., World Conference on Pavement and Asset Management, WCPAM2017, Milan, Italy – June 12/16, 2017.
5. Collective work: „Concrete Pavement Condition Assessment System", SOSN-B, General Directorate for National Roads and Highways (GDDKiA), Warsaw 2007.
6. Collective work: „Road pavement and pavement elements diagnostics", DSN, General Directorate for National Roads and Highways (GDDKiA), Warsaw 2015.

Analysis of selected input data impact on energy demand in office building - case study

Joanna Sinacka[1,*] and *Katarzyna* Ratajczak[1]

[1]Institute of Environmental Engineering, Poznan University of Technology, Berdychowo 4, 60–965 Poznan, Poland

Abstract. In the building designing process special attention should be paid to energy calculations - with analysis of different input data low energy or passive house can be obtained and decision of which parameters can or cannot be changed may be made. The following input data was analysed: shape factor A/V, glazed area, internal heat gains, air tightness, efficiency of heat recovery. The results show that the input data for the energy calculations are crucial for the obtained result.

1 Introduction

There are many parameters that need to be taking into account during building designing process. When planning to buy or build a building, an investor expects that the building will be esthetic, comfortable and practical in use, and the operating costs will be low. The primary goal of the designing process must be to meet the national legal requirements that apply to construction projects, but the next step, after fulfilling those requirements, should be to provide climate comfort at the lowest energy consumption, contributing to the low costs of maintaining, which are important to the user.

Energy efficiency of the buildings is regulated by the European Union. All new public buildings in Poland must be nearly zero energy buildings by 31st of December 2018 and others by 31st December 2020 [3,5]. In accordance with Polish regulations non-renewable primary energy ratio in new public buildings presently should be lower than 60 kWh/(m^2·a) and after 2020 year should decrease to 45 kWh/(m^2·a) [7].

The important aspects of energy efficient building designing are its shape and location in relation to the sides of the world. Well thought-out location of the windows, an appropriate selection of their size as well as the shading elements allow to employ a solar radiation (passive heat source) for heating. It is also important to provide the necessary air-tightness of the building as well as apply ventilation systems with heat recovery and building materials with good thermal insulation parameters to reduce heat exchange between the internal and external environment.

* Corresponding author: joanna.sinacka@put.poznan.pl

1.1 Passive House Standard

Passive House concept was defined by Passive House Institute of Darmstadt in Germany in the 1990s [1,2]. Development of Passive House Standard is consequence of the increasing requirements for thermal comfort and indoor air quality. The most important criterion is the energy demand for heating which should not exceed 15 kWh/(m^2·a) and the non-renewable primary energy which should be less than 120 kWh/(m^2·a) [2].

Passive House Standard (PH) characterise [4]: ensuring thermal transmission (U-value) for external walls below 0,15W/(m^2·K), no thermal bridges, high air tightness (≤0,6h^{-1}), ensuring high quality windows: low U-value (≤ 0,8W/m^2/K) and high total solar energy transmission factor (≥ 0,5), ventilation system with heat recovery efficiency ≥75%, limiting heat loss for domestic hot water installation, high efficiency of electricity use, non-renewable primary energy (nRPE)≤120 kWh/(m^2·a), heating demand ≤15 kWh/(m^2·a).

1.2 Polish legal requirements for buildings (Technical Conditions)

Polish regulations determine the thermal insulation of building envelope, as well as other requirements [7]:

a) Maximum U-values for building envelope by January 01st, 2017 – WT2017 (by December 31st, 2020 – WT2021): $U_{ExternalWall} \leq$ 0,23 (0,2) W/(m^2·K), $U_{Roof} \leq$ 0,18 (0,15) W/(m^2·K), $U_{GroundFloor} \leq$ 0,3 (0,3) W/(m^2·K), $U_{WINDOW} \leq$ 1,1 (0,9) W/(m^2·K), $U_{ExteriorDoor} \leq$ 1,5 (1,3) W/(m^2·K).
b) The maximum area of glazing – when windows have U-value higher than 0,9 W/(m^2·K), the total area of glazing should be lower than calculated with equation: $A_{W,MAX}=0,15 \cdot A_Z+0,03 \cdot A_W$, where A_Z is area of building plan in 5m wide next to external walls and A_W is remaining area of building plan in m^2.
c) Air flow for each person in office building should be not less than 20 m^3/h.
d) Natural ventilation can be used only if designed minimum air change per hour is not greater than 2 h^{-1}.
e) Air tightness in a building should be no less than n_{50}=1,5 h^{-1} for mechanical ventilation and no less than n_{50}=3,0 h^{-1} for natural ventilation.
f) When air flow is greater than 500 m^3/h, heat recovery system should be applied.
g) The maximum non-renewable primary energy index for heating, ventilation and domestic hot water preparation as well as for cooling and lighting depends on the type of building, and for office buildings from 31st December of 2020, it is respectively: 45 kWh/(m^2·a), 10 kWh/(m^2·a) (for cooling whole building), 50 kWh/(m^2·a).

2 Method

Analysis was conducted in design PH plug-in to Sketch Up program and in PHPP – passive house planning package, which can be used for designing energy efficient building as well [4]. The following input data were analyzed: shape factor A/V, glazed area, internal heat gains, air tightness, efficiency of heat recovery. Variants that meet the requirements of Polish technical conditions and requirements for passive buildings have been analyzed.

3 Building description

Analysis regards two-storey office building which is intended for 18 employees and its treated floor area is equal 303 m^2. Required air stream supply per person is 20 m^3/(h·p), so total supply air is equal 360 m^3/h. Building volume is equal 760m^3 so natural ventilation can be applied.

Constant input data used in the calculations: the interior temperature: 20 °C (winter) and 25°C (summer), the air stream supply: 360 m^3/h, heat recovery efficiency: 72,7%, the heating and domestic hot water (DHW) source: direct heating (efficiency of heating system: 0,87; efficiency of DHW system: 0,55), amount of DHW per person is 6 $dm^3/(d{\cdot}p)$ (energy demand for DWH 6,4 $kWh/(m^2{\cdot}a)$), the multi-split air conditioning system efficiency of 3,5.

In basic variants (PH, WT2021) windows area on the south side is equal to 15,33m^2, on the east side 14,58m^2, on the west side 18,0 m^2. On the north side windows were not designed. Building envelope for WT2021 and PH (in brackets) is characterise by means of U-value: $U_{ExternalWall}$=0,2 (0,1) $W/(m^2{\cdot}K)$, U_{Roof}=0,15 (0,1) $W/(m^2{\cdot}K)$, $U_{GroundFloor}$=0,3 (0,12) $W/(m^2{\cdot}K)$, U_{WINDOW} = 1,0 (0,8) $W/(m^2{\cdot}K)$, $U_{ExteriorDoor}$=1,3 (0,8) $W/(m^2{\cdot}K)$.

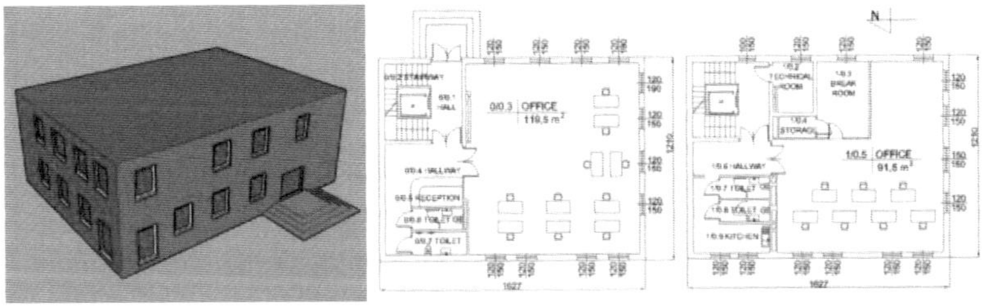

Fig. 1. Building envelope and building plan

4 Variants

Seven variants for passive house and building fulfilling Polish Technical Conditions will be analysed. For each variant heating and cooling purposes will be calculated. Comparison of parameters that will be changed in each variant are shown in Table 1.

Table 1. Calculation variants

Variant	Comparison of	Changing parameters	Purpose
0 basic	ventilation	natural/mechanical	if natural ventilation is enough to meet the requirements of the nRPE
1	shape factor	A/V [1/m] 0,66/0,82/0,92	impact of building shape
2	windows area	A_W [m^2] 25,72/51,07/134,2/175,16	if the building meets requirements of primary energy for heating and cooling
3	orientation	rotation of building by 0 / 90 / 180 / 270 °	the impact of building orientation on solar heat gains and consequently energy demand
4	internal heat gains	q_{int} [W/m^2] 3,5 / 5,7 / 6,2 / 8,8	choosing proper values from many guidelines [6.8]
5	air tightness	n_{50} [h^{-1}] 0,6/1/1,5/3	impact of air tightness on energy need
6	efficiency of heat recovery	η_{HE} [%] 50/60/70/80/90	impact of efficiency of heat recovery on heat losses

5 Results and discussion

5.1 Variant 0– Basic variant

Calculations of heating and cooling energy need for the building were made taking into account input data from section 3. The results are shown on figure 2.

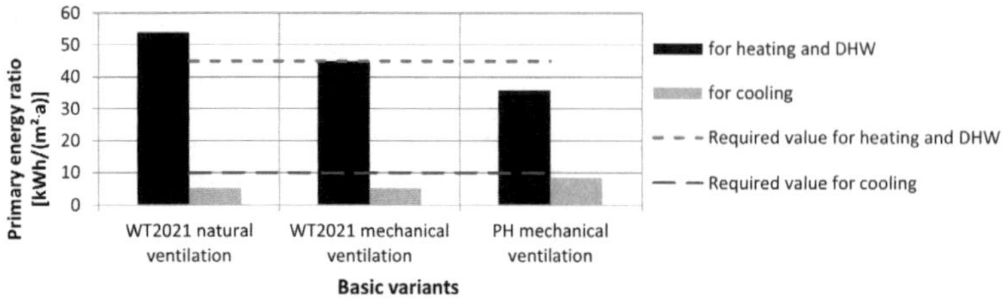

Fig. 2. Basic variants: WT2021 with natural ventilation, WT2021 with mechanical ventilation, PH

Despite applying all data that are required in polish building codes for the year of 2021 non-renewable primary energy criterion is not achieved for building with natural ventilation. Mechanical ventilation with heat exchanger for this building is necessary. This will be basic variant (WT2021) for all other calculations and the other results will relate to it. To show what is the impact of the same input data for passive house all results will relate to passive house (PH) as well.

5.2 Variant 1 – impact of the building shape factor A/V on energy demand

The building shape factor A/V is defined as a quotient of sum of the external areas of building envelope and the volume of building envelope. The higher A/V, the higher is the area of surface through which heat flow. Energy efficient buildings must have as low A/V as possible. In calculations three different shapes of the building with the same treated floor area and with windows on the same orientation were compared. The difference is in the area of walls, roof and ground floor. All remaining parameters of the building were unchanged. On figure 3 building shapes are shown. The results of impact of shape factor on energy need for heating and cooling are presented on figure 4.

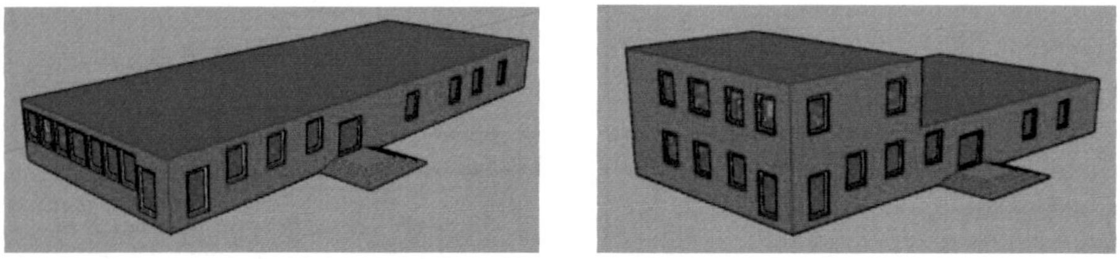

Fig. 3. Building with shape factor of A/V=0,77 m^{-1} (left) and A/V=0,87 m^{-1} (right)

The shape factor of analysed buildings are for basic variant A/V=0,61m^{-1} (two-storey building) , 0,77 m^{-1} for one storey building and 0,87 m^{-1} for one and a half storey building.

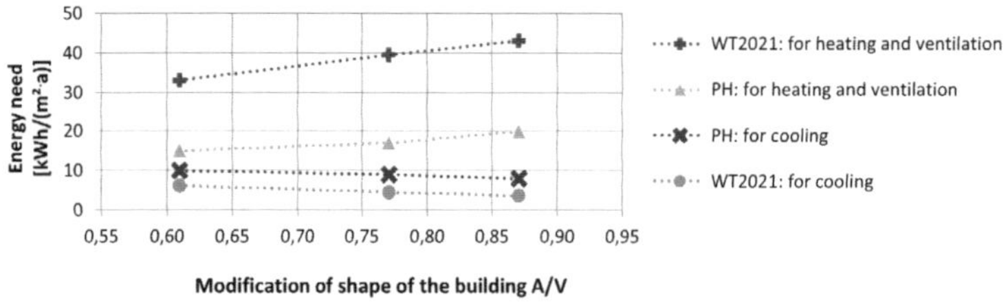

Fig. 4. Impact of the building shape factor A/V on heating and cooling energy need

The results show that energy need for heating rises with A/V factor. This is due to the increase in surface area through which heat is lost. At the same time energy need for cooling decreases.

5.3 Variant 2 – impact of the designed windows area in the building

On figure 5 results of calculations of non-renewable primary energy for heating and cooling with maximal value are shown. Changing glazed area in the building from 25 to 175 m^2 doesn't affect much energy for heating need. Windows have higher heat transfer coefficients, so the building's heat losses are higher, but at the same time more solar heat gains penetrate the building, so heat balance is not changed.

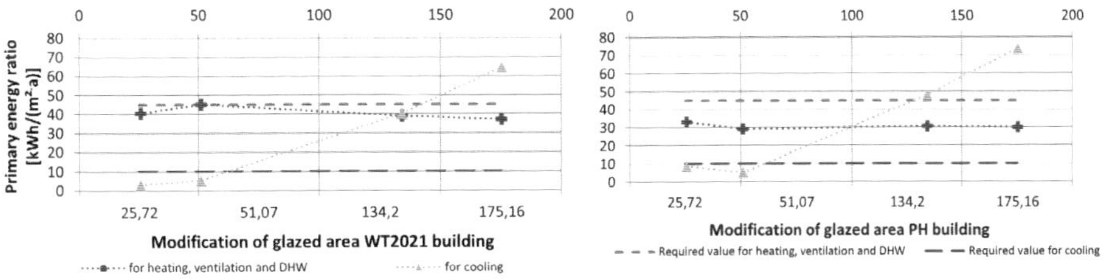

Fig. 5. Impact of glazed area on heating and cooling energy demand

Despite value of energy for heating does not vary much with different glazed area, those areas affects energy for cooling. With increasing glazed area energy for cooling need increases rapidly. Maximum value of non-renewable primary energy ratio for cooling is not exceeded for windows area limited with polish requirement, even when for windows with very low heat transfer coefficient (0,8 $W/(m^2 \cdot K)$. When designing buildings with huge glazed area special attention should be paid to shading devices, which can decrease energy need for cooling.

5.4 Variant 3 – impact of the building orientation

In the variant 3 the building is rotated up 90°, 180° and 270°. Solar heat gains (SHG) depend on building orientation but they are the same for PH and WT2021: for +90° solar heat gains are equal 29,73 kWh/m^2/a, for +180° SHG =22,25 kWh/m^2/a, for +270° SHG=32,87 kWh/m^2/a. This analysis show how important is designing a functional layout in such a way that windows on the north side are limited, whereas area of windows on the south side is possibly highest. On the other hand with glazed area on the south side the energy for cooling increases, so designer should find compromise between decreasing energy for heating with no increase of energy for cooling.

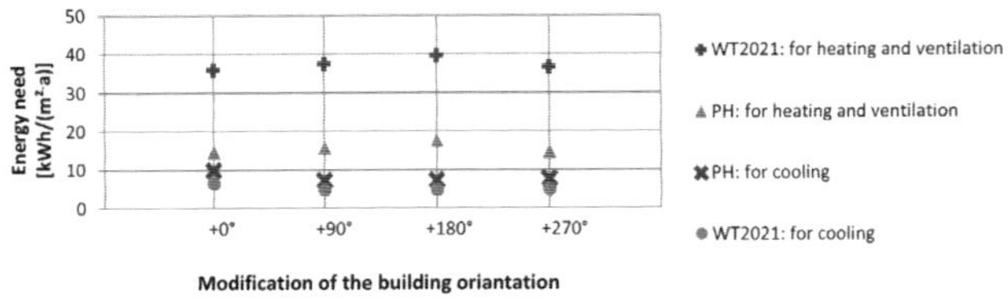

Fig. 6. Impact of building orientation on useable energy

Impact of building orientation on useable energy is presented on figure 6. The highest heating demand is when building is rotated up 180° and windows area on north side is the highest. The highest difference is equal to 2,9kWh/m^2/a for PH and 3,8kWh/m^2/a for WT2021.

5.5 Variant 4 – impact of the internal heat gains

The amount of internal heat gains that are used in calculations should be carefully selected. Different methodologies for calculating energy demand provide various internal heat gains indicators. Internal heat gains (IHG) have significant influence on energy demand. Each calculated building should be analysed in this respect in great detail and in accordance with its equipment. Incorrect assumptions lead to uncomfortable working conditions. Oversize IHG cause building unheated in winter time, whereas undersize IHG cause an overheating in summer time. Boost q_{int} from 3,5W/m^2 to 8,8W/m^2 for PH building causes greater increase in energy demand for cooling than the decrease in heating demand (fig. 7).

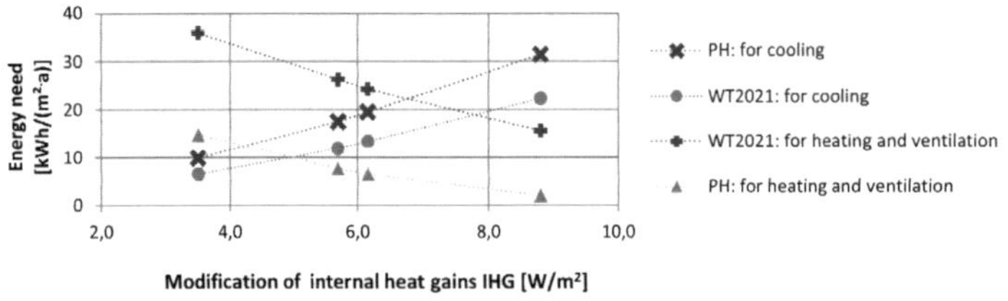

Fig. 7. Impact of internal heat gains (IHG) on energy demand for heating and cooling.

5.6 Variant 5 – impact of the building air tightness

In variant 5 ventilation system is analysed. Heat losses rate is equal H_{VE}=22,7 W/K for each case, whereas transmission heat losses rate is equal 106,1W/K for PH and 165,9 W/K for WT2021. Heat losses by infiltration depends on air change rate at press. test but is equal for PH and WT2021 (for n_{50}=0,6h^{-1} H_{INF} =11,5 W/K), for n_{50}=1,0h^{-1} H_{INF}=19,3W/K, n_{50}=1,5h^{-1} H_{INF}=29W/K, n_{50}=3,0h^{-1} H_{INF}=57,8W/K.

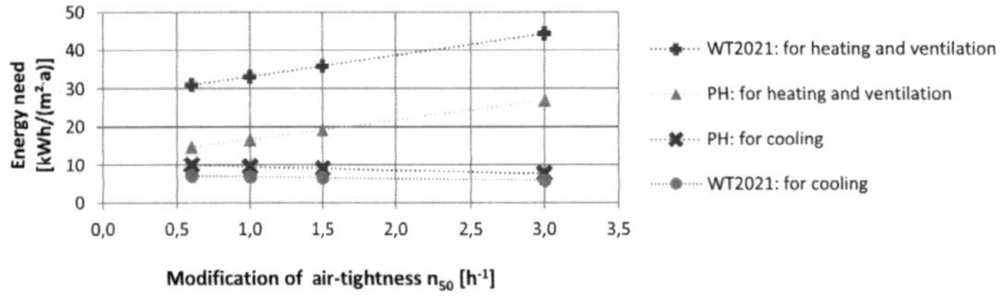

Fig. 8. Impact of air change rate (n_{50}) on energy demand for heating and cooling.

When H_{inf}/H_{ve} ratio increases the importance of mechanical ventilation with heat exchanger decreases and air flow is not under control (for analysed building if n_{50}=0,6h^{-1} H_{inf}/H_{ve} ratio is equal 34% but for n_{50}=3,0h^{-1} increases to 72%). In addition, an increase of n_{50} about 0.9 h^{-1} presents heating energy demand an increase about 5kWh/m^2/a for WT2021 and about 4,5kWh/m^2/a for PH. Cooling energy demand is more stable and for WT2021 decreases about 0,5 kWh/m^2/a and for PH decrease about 1 kWh/m^2/a. On the graph (figure 8) the impact of air change rate (n_{50}) on energy need for heating and cooling is shown. Ensuring proper (high) air-tightness especially in low energy hand passive houses is very important.

5.7 Variant 6 – impact of the effective heat recovery efficiency (ventilation system)

An effective and economical way to supply a fresh air for buildings is mechanical ventilation with heat recovery. In Variant 6 infiltration heat losses rate are equal 11,5W/K for PH and 29W/K for WT2021. Heat losses by ventilation system H_{vs} for heating depends on effective heat recovery efficiency $\eta_{HR,eff}$ and is the same for PH and WT2021. For $\eta_{HR,eff}$=50% H_{vs}=41,6W/K, for $\eta_{HR,eff}$ =60% H_{vs}=33,3W/K, for $\eta_{HR,eff}$ =65% H_{vs}=29,1W/K, for $\eta_{HR,eff}$=70% H_{vs}=25W/K, for $\eta_{HR,eff}$=75% H_{vs}=20,8W/K, for $\eta_{HR,eff}$=80% H_{vs}=16,7W/K, for $\eta_{HR,eff}$=85% H_{vs}=12,5W/K, for $\eta_{HR,eff}$=90% H_{vs}=8,4W/K. In case of cooling (summer time) there is no heat recovery. Mechanical ventilation is realized by bypass.

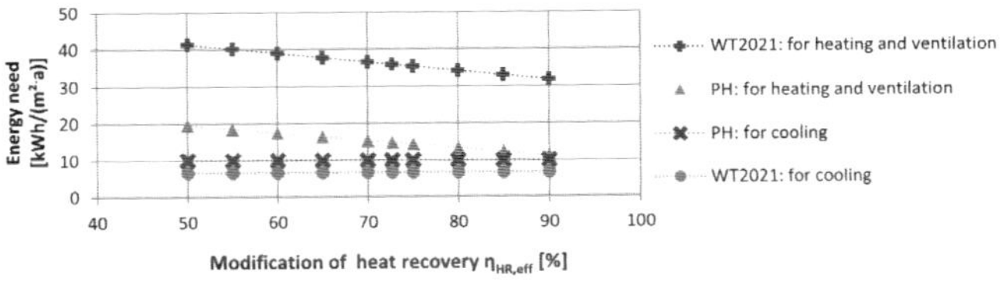

Fig. 9. Useable energy and heat losses by ventilation system versus total ventilation heat losses ratio

Analysis, which results are presented in Fig. 9, confirm the importance of effective heat recovery efficiency for an useable energy. An increase of efficiency $\eta_{HR,eff}$ leads a decrease energy demand, for example increase of $\eta_{HR,eff}$ about 20% presents heating energy demand a decrease about 4,8kWh/m^2/a for WT2021 and about 4,0 kWh/m^2/a for PH. High efficiency $\eta_{HR,eff}$ is necessary for energy-efficient building.

It is important to provide mechanical ventilation with heat recovery, because with no heat recovery it is not possible to meet requirements for non-renewable primary energy and low energy demand for heating, especially, for passive house standard.

6 Summary

The results showed that input data for energy calculations are very important, because they can change the final result and they are crucial to meet the requirements of energy standards. Designer can decide what is the most important factor in the design and choose appropriate solution to receive required building parameters. For example, when building has large area of glazing which is characterized by a high heat transfer coefficient and it influences significantly the high usable energy, reducing energy consumption is possible by using other components with better parameters (e.g. mechanical ventilation system with higher heat recovery efficiency). Another example is a building with a complicated shape so its shape factor is far from perfection. In order to reduce energy consumption building materials with a lower thermal conductivity can be used.

Simulations of different variants which take into account various input data can be very helpful in the design process, while knowledge of impact the various input data on the final result allows to choose design variants properly.

When designing a building, the final effect should be considered. Of course, the precedent requirements are those resulting from the legal regulations and standards, however building that meets these requirements may have different levels of energy consumption for heating and cooling. Modeling of energy demand and analysis of the various input data impact on the final result are helpful during decision making process: how a building should look like, what materials and equipment should be applied and what will be the energy consumption - meaning the costs of the building and operation costs.

References

1. Ł. Amanowicz, E. Szczechowiak, COW, 48, (2017)
2. V. Badescu, N. Rotar, S. Budea, Sustainable Cities and Society 24, (2016)
3. V. Badescu, N. Rotar, I. Udrea, Energ. Effic. 8, (2015)
4. Directive 2010/31/EU ,Official Journal of the European Union L 153/13
5. W. Feist, PHPP Passive House Planning Package – The energy balance and design tool for efficient buildings and retrofits (Passive House Institute, Darmstadt, Germany, 2015)
6. M. Kassai, , Energy Procedia, 112 (2017)
7. Polish law: Dz.U. poz. 376, (2015)
8. Polish law: Dz.U. poz. 1422, (2015)
9. VDI 2078:2015-06

Assessment of temperature susceptibility for rubber granulate modified road asphalt binders considering impact of aging

Mieczysław Słowik[1*], *Damian* Wiśniewski[2], *Marcin* Bilski[1], *Marta* Mielczarek[1]

[1]Poznan University of Technology, Faculty of Civil and Environmental Engineering, Institute of Civil Engineering, Piotrowo 5, 61-138 Poznań, Poland
[2]UTP University of Science and Technology in Bydgoszcz, Faculty of Building, Architecture and Environmental Engineering, Al. Prof. S. Kaliskiego 7, 85-796 Bydgoszcz, Poland

Abstract. The article presents the results of research on temperature sensitivity of ground rubber modified asphalt binders. The tests were performed for 50/70 penetration grade road bitumen and two kinds of granulated rubber: activated and inactivated. Such parameters of asphalt binders as: penetration, softening point, dynamic shear modulus and phase angle have been determined and on this basis the temperature susceptibility of the reference bitumen and asphalt binders modified by activated and inactivated rubber have been determined. All the tests were carried out for three observation levels: without aging and after aging simulations carried out by RTFOT and PAV methods. The tests revealed that addition of granulated rubber has a positive influence on asphalt binder properties, even after the aging process, causing an increase in the softening point and a decrease in penetration and the phase angle of asphalt binder. Asphalt binders obtained in result of modification by addition of rubber show smaller temperature susceptibility than the reference bitumen (without addition of rubber). Also, a positive impact of aging on the temperature susceptibility was observed for the investigated binders.

1. Introduction

Nowadays, more and more attention is being paid to ecological solutions (protecting the environment). It also applies to road engineering whose research centers search for new materials and trying to identify development directions for material-technological solutions which take into account the rules of sustainable development [1]. One of additives used for production of asphalt mixtures is crumb rubber recovered from used car tires. Rubber scrap is very harmful to the natural environment due to its very long degradation period, reaching even more than 100 years. Therefore, its reusing is a very good utilization method, especially that, due to the development of motorization industry, the number of tires is constantly increasing and subsequently, rubber scrap needs to be managed in an ecological manner. Thus, the issues raised in our article fit into the subject of sustainable development.

* Corresponding author: mieczyslaw.slowik@put.poznan.pl

One of the materials most commonly used in road pavements is bitumen which blended with aggregate and filler at high temperature makes up a Hot Mix Asphalt (HMA). It is supposed to bind all the components of asphalt mixture. Bitumen belongs to thermorheological materials. In result of heating, bitumen gradually changes from solid to liquid state which leads to viscosity reduction. However, when cooled, it changes from liquid to solid state which is accompanied by an increase in its viscosity and stiffness modulus. The type of asphalt binder and its properties have a large influence on properties of the asphalt mixture which, subsequently affects performance of the road pavement.

The properties of road bitumen depend on temperature and time of loading. Therefore, it is necessary to increase the viscoelastic range of asphalt binder to maintain its rheological state, despite extremal temperatures reached by the pavement during service. A wide range of viscoelasticity can be obtained through the bitumen modification which involves adding appropriate amounts of modifying agents which include: fibers (polyester, polypropylene, mineral wool, cellulose), sulfur, natural asphalts (e.g.. Trynidad Epuré, Gilsonite), fillers, organometallic compounds, polymers, or synthetic wax.

By using modifying additives in asphalt binder it is possible to obtain the following effects: decreasing temperature susceptibility [1÷8], increasing viscosity [6,9], increasing dynamic shear modulus at high temperatures and its decreasing at low temperatures [2,4,5,8], reduction of Fraass breaking point [7], raising the softening point [6,8,9], penetration value change [6,9].

Used tire rubber is considered to be an asphalt binder modifier. The obtained ground rubber can be added directly to bitumen (wet 'modification') or to asphalt mixture, through replacement of a part of fraction of aggregate ('dry' modification). Rubber crump addition by the 'wet' method or the 'dry' one has a positive influence on different properties of asphalt binder as well as asphalt mixture, which is confirmed by a great number of tests [4,5,6,7,8,9,10,11,12,13,14]. The influence depends on the degree of fineness and the amount of used granulated rubber which affect the bitumen viscosity, its rheological properties, softening point change, the binder penetration and values of dynamic shear modulus [15,16,17,18].

2. Research subject

The aim of the study was to assess the influence of road bitumen selected properties, especially on its temperature susceptibility and resistance to aging. 50/70 penetration grade bitumen, commonly applied in asphalt pavements, was used for the tests. The tests were performed for three kinds of binder: reference bitumen 50/70 (with no addition of rubber), asphalt binder 50/70 with addition of 15% (m/m) activated rubber and asphalt binder 50/70 with addition of 15% (m/m) of inactivated rubber.

The process of activation was performed with the use of a production line consisting of the following elements: electromagnetic mill with a generator of microwaves, centrifugal rotor turning with the supersonic speed and other auxiliary equipment. The ground car tire rubber was successively processed, initially by means of mechanical- chemical activation, next it was subjected to synthesis and nano-synthesis. It allowed to obtain a new material whose structure is very similar to polymer structure [19].

All binders were tested for their rheological and functional properties at three observation levels: unaged, subjected to short term aging and then long term aging. On the basis of the carried out tests assessment of temperature susceptibility was performed for unmodified bitumen and crump rubber modified asphalt binders.

3. Asphalt binders research methodology

3.1. Determination of bitumen penetration and softening point

Penetration and softening temperature belong to the basic properties of asphalt binders. Penetration is the measure of consistency (stiffness) of the binder. Softening point value defines the conventional, upper boundary of the viscoelastic state.

Determination of penetration with the use of a needle is performed according to the standard PN-EN 1426:2015-08 Bitumen and bituminous binders. Determination of needle penetration. The test involves determination of depth (in 0.1 mm), a penetration needle with mass 100 g enters a specimen of bitumen in time of 5 s at 25°C.

Determination of softening point is carried out according to Ring and Ball method described in the standard PN-EN 1427:2015-08 Bitumen and bituminous binders - Determination of the softening point - Ring and ball method.

3.2. Determination of rheological properties of asphalt binders by Dynamic Shear Rheometer (DSR) method

The test in a rheometer DSR is carried out in order to define complex rheological properties of asphalt binders – viscoelastic materials. The test involves determination of complex shear modulus G^* and phase angle δ.

Value G^* is expressed as a ratio of maximal value of shear stress to the maximal value of strain. Value δ is a difference of phases between strain and stress in case of harmonic vibrations occurrence. When δ=90°, then asphalt binder can be considered an ideally viscous liquid, whereas when δ=0°, the material is found to be an ideally elastic body [20]. It is assumed that between these two boundary values, asphalt binder exhibits viscoelastic properties. The above values can be defined in the following way:

$$|G^*| = \sqrt{G'^2 + G''^2} \tag{3.2}$$

or as a complex number:

$$G^* = G' + iG'' \tag{3.2}$$

and

$$\delta = \mathrm{tg}^{-1}\left(\frac{G''}{G'}\right) \tag{3.3}$$

where:

$G' = |G^*| \cos \delta$ - is a storage modulus, which defines the material ability to store energy (real part of complex shear modulus);

$G'' = |G^*| \sin \delta$ – is a loss modulus which defines the material ability to dissipate energy (imaginary part of complex shear modulus);

$tg\delta = \frac{G''}{G'}$ is a damping factor;

i – is an imaginary unit described by the equation $i^2 = -1$;

$|G^*|$ - dynamic shear modulus;

G^* - complex shear modulus

δ –phase angle.

The test in Dynamic Shear Rheometer was carried out according to the standard PN-EN 14770:2012 Bitumen and bituminous binders - Determination of complex shear modulus and phase angle using a Dynamic Shear Rheometer (DSR). The test procedure involved placing an asphalt binder specimen with mass of 0.100÷0.105 g on a measurement head (plate) with diameter of 8 mm., which was thermostated for 10 min. in the temperature of testing. After thermostating process a measurement was started, for constant angular frequency equal to 10 rad/s and in a fixed temperature, starting with 82˚C, and next gradually (every 1˚C) down to 0˚C.

3.3. Test of asphalt binder resistance to aging

The asphalt binders were also subjected to aging: short term aging simulated in accordance with the standard PN-EN 12607-1:2014-12 Bitumen and bituminous binders. Determination of the resistance to hardening under influence of heat and air. RTFOT method, and next long term aging simulated according to the standard PN-EN 14769:2012 Bitumen and bituminous binders - Accelerated long-term ageing conditioning by a Pressure Ageing Vessel (PAV).

RTFOT (Rolling Thin Film Over Test) method involves putting a thin film of an asphalt binder under the impact of heated air at a period of time equal 75±1 min. Bitumen specimen is poured to glass bottles which are successively placed in a turning disc turning with the speed 15,0±0,2 rpm, placed in a laboratory oven and exposed to temperature 163˚C ± 1˚C and compressed air with flow rate 4,0 l/min ± 0,2 l/min and maximum pressure 200 kPa. RTFOT method simulates changes within the structure and properties of asphalt binder occurring during technological processes of asphalt mixture production and paving (mixing the aggregate with bitumen, transport of asphalt mixture, paving and compacting of the asphalt layer).

PAV (Pressure Aging Vessel) method involves exposing a specimen of the binder to the impact of air under pressure 2.1 MPa ± 0.1 MPa in time of 20 h at 90˚C, 100˚C or 110˚C, sustained with tolerance of ± 0,2˚C. The asphalt binder tested by this method must first be subjected to aging by RTFOT. In the tests performed as part of this study the binder resistance to aging was assessed by means of PAV method in temperature 100±0,2˚C. PAV method simulates changes that occur in the asphalt binder during long term operation of the asphalt pavement (app. 10. years).

3.4. Determination of values characteristic of asphalt binder temperature susceptibility

Temperature susceptibility of a binder is commonly defined by means of penetration index which is determined on the basis of the asphalt binder softening point and penetration. Penetration index specifies how the asphalt binder changes its consistency (stiffness) in relation to temperature changes. The higher value of the penetration index the lower asphalt susceptibility to temperature changes. Penetration index (PI) is determined using the following formula [21]:

$$PI = \frac{1952 - 500\log(Pen_{25}) - 20T_{PiK}}{50\log(Pen_{25}) - T_{PiK} - 120} \tag{3.1}$$

where:

PI – penetration index, [-];
Pen_{25} – penetration of a binder at 25˚C, [mm/10];
T_{PiK} – softening point of a binder, [˚C].

Temperature susceptibility of the binder can also be defined by means of SMI - Shear Modulus Index. This index is determined from the value of dynamic shear modulus obtained using Dynamic Shear Rheometer for a wide range of temperatures that occur during operation of road pavements (in this case from 82°C to 0°C), much wider than for determination of the penetration index. It enables more accurate assessment of temperature susceptibility. SMI index is determined on the basis of Walther's formula [2]:

$$SMI_{T_2/T_1} = \left| \frac{\log\log\left|G^*_{T_1}\right| - \log\log\left|G^*_{T_2}\right|}{\log(T_1 + 273{,}15) - \log(T_2 + 273{,}15)} \right| \tag{3.2}$$

where:

SMI – Shear Modulus Index (describing the temperature susceptibility of the binder)

$\left|G^*_{T_1}\right|$ - dynamic shear modulus at T_1, [Pa]

$\left|G^*_{T_2}\right|$ - dynamic shear modulus at T_2, [Pa]

$T_1; T_2$ – initial and final temperature in DSR tests ($T_1 > T_2$), [°C]; T_1=82°C, T_2=0°C was assumed.

4. Analysis of results

4.1. Penetration Index PI

The results of penetration and softening point of the analyzed asphalt binders are presented in table 1, whereas table 2 includes values of asphalt binders penetration index. The tests results are presented as the arithmetic mean values and uncertainty interval determined at significance level α=0.05 according to the methodology described in [22].

Addition of rubber granulate, both activated and inactivated, causes reduction of asphalt binder penetration. Activated rubber modified asphalt binder exhibits higher toughness as compared to inactivated rubber modified one. Penetration decrease is so big that there follows a shift of bitumen in classification taking into account penetration at 25°C - from 50/70 to 35/50. After RTFOT aging the asphalt binders modified with activated rubber revealed penetration decrease, whereas, for binders modified with inactivated rubber a decrease in penetration was reported. As regards RTFOT and PAV aging methods, it was observed that modification with rubber (activated and inactivated) caused an increase in binder penetration. However, changes in the value of penetration for asphalt binders subjected to the aging processes are significantly smaller than before aging.

The influence of granulated rubber is also observed in the case of softening point values. Addition of 15% of rubber to the bitumen increased the softening point, whereas, the highest increase was reported for inactivated rubber – by app. 19% as compared to the reference bitumen (unaffected with the process of aging). The same dependency was found for asphalt binder affected by aging RTFOT and PAV (increase by app. 12% as compared to the reference bitumen after the process of aging RTFOT and PAV). Softening point increase was also observed for asphalt binder exposed to aging RTFOT, however, the increase is higher than before and after aging RTFOT and PAV. It needs to be noted that, in this case, a bigger increase in the softening point was obtained after adding activated rubber to bitumen (softening point 14°C higher than for the reference bitumen after RTFOT).

Addition of granulated rubber (both activated and inactivated) has a positive impact on asphalt binder temperature susceptibility through increasing the value of penetration index for all observation levels – before and after aging RTFOT and RTFOT + PAV, as compared to the reference bitumen. The process of aging, both RTFOT and RTFOT+PAV caused an increase in the penetration index for all the analyzed kinds of asphalt binders – the reference one and activated and inactivated rubber modified ones. The highest values of penetration index were reported for the tested asphalt binders after RTFOT and PAV aging.

Table 1. Results of determination of penetration and softening point of the investigated asphalt binders

Asphalt binder	Penetration at 25°C [mm/10]			Softening point [°C]		
	Before aging	after RTFOT	after RTFOT +PAV	Before aging	after RTFOT	after RTFOT +PAV
50/70	56,9± 0,3	36,6± 0,5	20,4± 0,7	49,3± 0,2	54,6± 0,3	63,4± 0,5
50/70 + 15% ACTIVATED RUBBER	41,9± 1,0	35,8± 0,7	22,8± 0,8	55,9± 0,5	68,6± 0,2	68,6± 0,4
50/70 + 15% INACTIVATED RUBBER	46,6± 0,6	38,2± 0,8	25,1± 0,5	58,9± 0,2	62,6± 0,3	71,4± 0,4

The highest values of penetration index were reached for inactivated rubber modified asphalt binders – this effect was achieved for all observation levels. Before the process of aging, PI value for inactivated rubber modified asphalt binder increased by 1.72, whereas, for activated rubber modified binder, the value increased by 0.84, as compared to the reference one. After RTFOT aging the non-activated rubber modified asphalt binder reached PI value higher by 1.70 than the reference one after RTFOT aging (for activated rubber modified asphalt binder the value of PI was by 0.78 higher than for the reference one after RTFOT). After RTFOT and PAV aging PI value of reactive rubber modified asphalt binder increased by 1,69, whereas for activated rubber modified one it increased by 1.06 in comparison with the reference bitumen after RTFOT and PAV aging.

Table 2. Values of the tested asphalt binder penetration index

Asphalt binder	Penetration index		
	Before aging	After RTFOT	After RTFOT+PAV
50/70	-1,08±0,04	-0,82±0,04	-0,25±0,07
50/70 + 15% ACITIVATED RUBBER	-0,24±0,09	-0,04±0,04	0,81±0,06
50/70 + 15% INACTIVATED RUBBER	0,64±0,03	0,88±0,05	1,44±0,05

4.2. Shear Modulus Index SMI

Figures 1-6 show the results of asphalt binders tests carried out for a wide range of temperatures (from 0°C to 82°C) by the method of Dynamic Shear Rheometer (DSR). Figures 1-3 show diagrams of the dynamic shear modulus $|G^*|$ in relation to temperature, whereas, figures 4-6 depict diagrams of phase angle δ dependency, in relation to temperature. It can be observed, on the basis of the tests results, that the diagrams of dependence $|G^*| = f(T)$ and $\delta = f(T)$ show significant differences in the modified asphalt binders properties, as compared to bitumen without addition of rubber, and slight differences between properties of activated and inactivated rubber modified asphalt binders. The highest inclination of diagram $|G^*| = f(T)$ has been observed for the reference 50/70 penetration grade bitumen, whereas, for rubber modified asphalt binders the inclination decreases. It confirms the conclusion that addition of rubber decreases temperature susceptibility of asphalt binders. Modification of bitumen with rubber has also a positive influence on the phase angle which has significantly decreased, especially in the range of high surface operation temperature (especially above 30°C). It reflects improvement in elastic properties of asphalt binders through their modification with addition of granulated rubber.

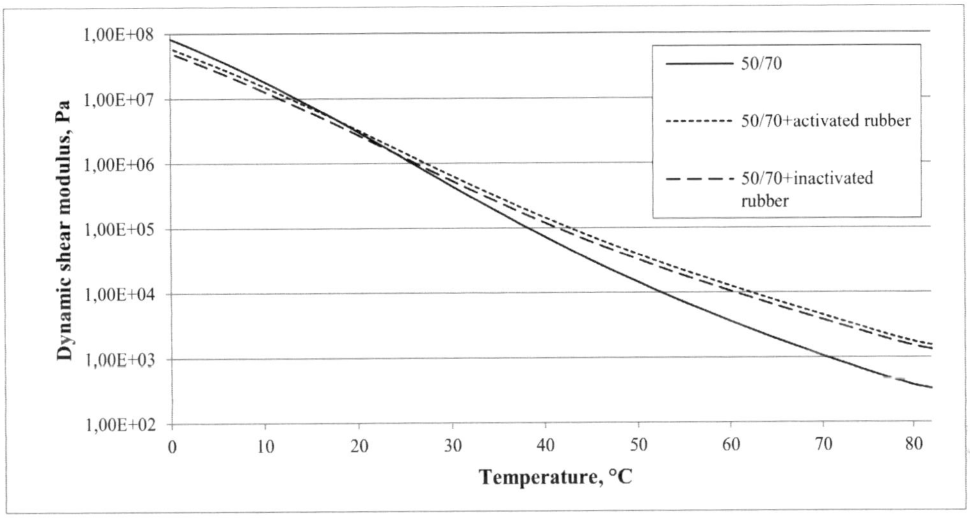

Fig. 1. Dynamic shear modulus in relation to temperature of unaged asphalt binders

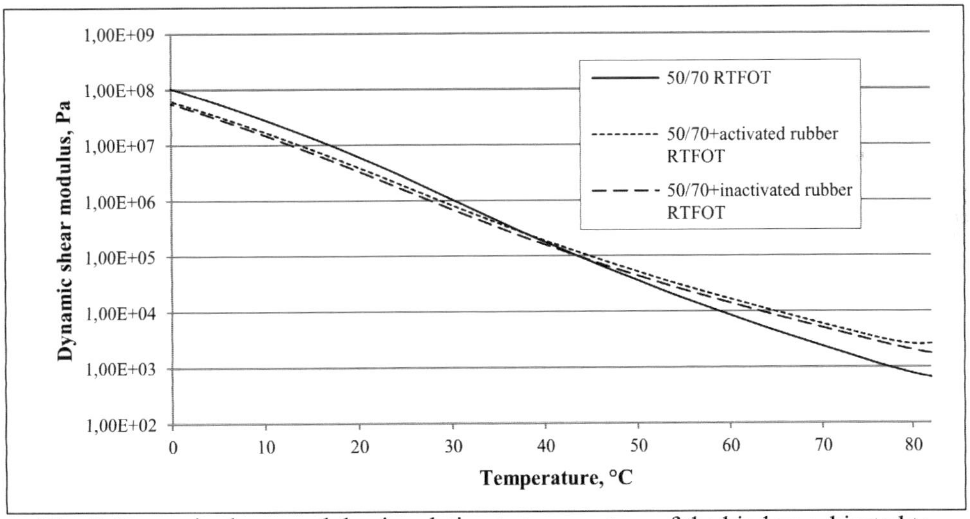

Fig. 2. Dynamic shear modulus in relation to temperature of the binders subjected to RTFOT aging

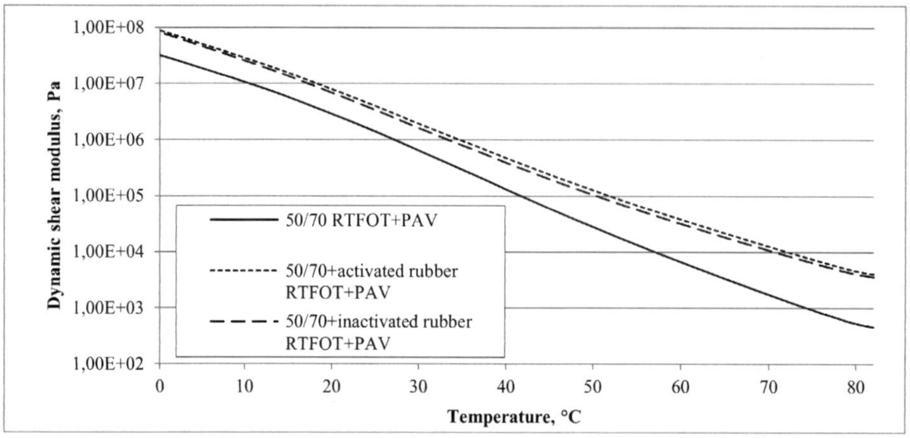

Fig. 3. Dynamic shear modulus in relation to temperature of the binders subjected to RTFOT and PAV aging

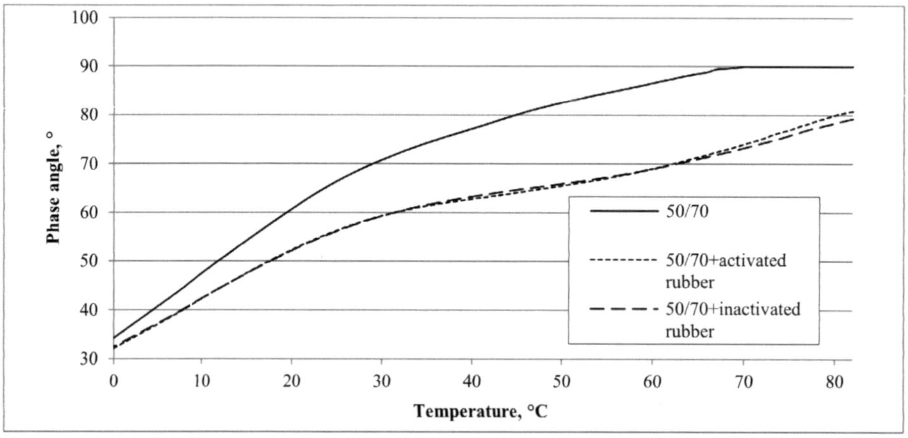

Fig. 4. Phase angle in relation to temperature of unaged asphalt binders

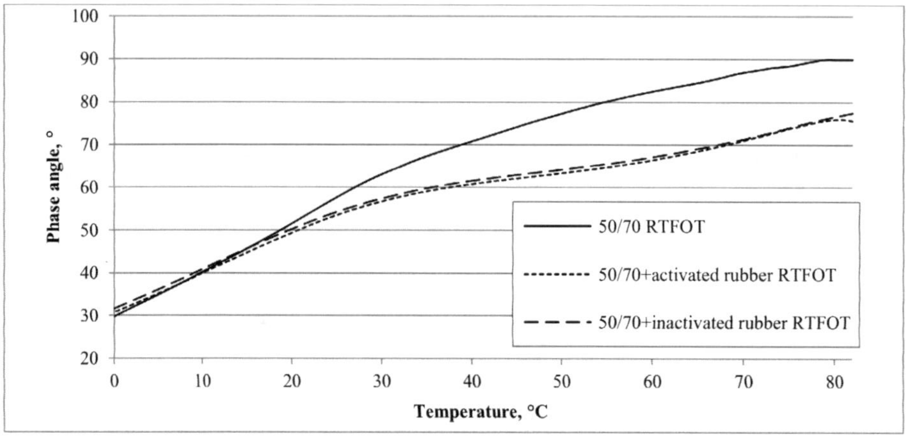

Fig. 5. Phase angle in relation to temperature of the binders subjected to RTFOT aging

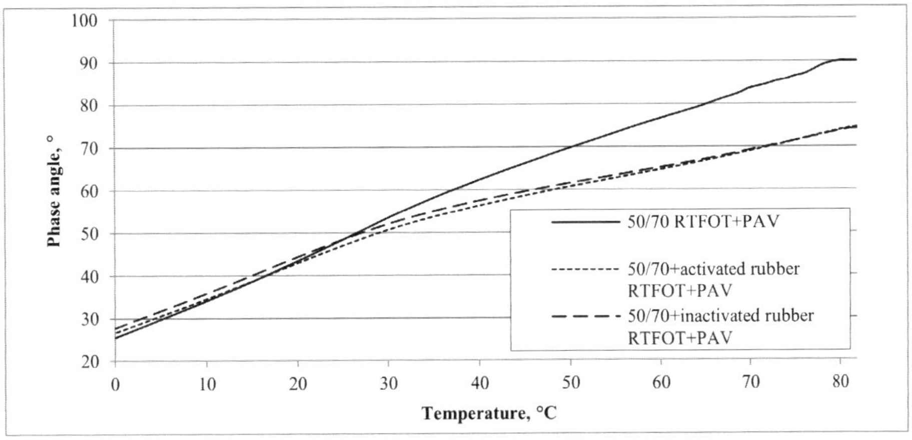

Fig. 6. Phase angle in relation to temperature of the binders subjected to RTFOT and PAV aging

Values of the shear modulus index SMI are presented in table 3. Addition of granulated rubber, both activated and inactivated, had an impact on reduction of SMI values, as compared to the reference bitumen, which proves its positive influence on asphalt binder temperature susceptibility. The lowest SMI values were observed for inactivated rubber modified asphalt binder both before and after short term and long term aging. As for activated rubber modified asphalt binder, lower SMI values before and after aging, were observed than for the reference bitumen which shows an improvement in the temperature susceptibility, slightly higher for inactivated rubber modified asphalt binder (4-8% as compared to inactivated rubber modified asphalt binder).

On the basis of carried out tests it was observed that the process of aging has a positive influence on asphalt binder temperature susceptibility. The values of the SMI were found to be lower after aging than those before aging, for all the investigated binders. The lowest values of the shear modulus index (for the reference bitumen, as well as for activated and inactivated rubber modified asphalt binders) were reported for binders after RTFOT aging, though, SMI values after RTFOT and PAV aging are only slightly higher than after RTFOT aging (maximally 0.5%).

Table 3. Values of shear modulus index of the investigated asphalt binders

Asphalt binders	$SMI_{0/82}$		
	Before aging	After RTFOT	After RTFOT+PAV
50/70	4,36	3,40	3,44
50/70 + 15% ACTIVATED RUBBER	3,98	3,14	3,31
50/70 + 15% INACTIVATED RUBBER	3,94	3,01	3,04

5. Conclusions

The following conclusions can be formulated on the basis of the results of tests performed for activated and inactivated rubber modified asphalt binders.

a) Addition of granulated rubber (both activated and inactivated) has an influence on raising the softening point and reduction of 50/70 bitumen penetration, which causes a shift of the bitumen in classification concerning penetration determined at

25°C from type 50/70 to 35/50, which means that in result of rubber addition, bitumen has been stiffened. Similar changes in asphalt binder stiffness can be observed on the basis of analysis of the impact of short and long term aging, however, in this case the changes are smaller than for unaged bitumen.

b) Addition of rubber to 50/70 penetration grade bitumen causes an increase in the value of the analyzed binder penetration index, both before and after short and long time aging, whereas the lowest values of penetration index was found for the reference bitumen, whereas the highest for inactivated rubber modified asphalt binder. This proves that addition of rubber has a positive impact on a decrease in temperature susceptibility of the analyzed binder which is a desirable effect.

c) The process of aging improves temperature susceptibility of investigated asphalt binders through increasing penetration index PI (the highest values were obtained for asphalt binders after RTFOT and PAV aging) and shear modulus index SMI (the highest values were found for asphalt binders after RTFOT aging).

d) Addition of 15% of rubber (activated and inactivated) leads to reduction of the dynamic shear modulus value at low temperatures and its increase at high road pavement operational temperatures which confirms a positive impact of the rubber addition on asphalt binders temperature susceptibility.

e) Addition of activated and inactivated rubber contributes to a decrease in asphalt binder phase angle. This decrease is most visible for the high pavement operational temperatures (over 30°C). On this basis, it can be concluded that addition of rubber causes improvement of the binder elastic properties.

Literatura

1. Radziszewski P., Nazarko J., Vilutienė T., Dębkowska K., Ejdys J., Gudanowska A., Halicka K., Kilon J., Kononiuk A., Kowalski K., Król J., Nazarko L., Sarnowski M.; Future trends in road pavement technologies development in the context of environmental protection; The Baltic Journal of Road and Bridge Engineering; 11(2):160-168, (2016).
2. Słowik M.; Thermorheological Properties Of Styrene-Butadiene-Styrene (SBS) Copolymer Modified Road Bitumen, Procedia Engineering 208: 145-150 (2017).
3. Lu X., Isacsson U.; Chemical and rheological evaluation of ageing properties of SBS polymer modified bitumens; Fuel 77(9): 961-972 (1998).
4. Navarro F.J., Partal P., Martinez-Boza F., Gallegos C.; Thermo-rheological behavior and storage stability of ground tire rubber-modified bitumens; Fuel 83(14): 20412019 (2004).
5. Sybilski D., Bańkowski W., Horodecka R., Wróbel A., Mirski K.; Method for modification of mineral-asphalt with addition of rubber „tecRoad"; Drogownictwo (6): 189-193 (2011).
6. Koba H., Skotnicki Ł., Szydło A.; Properties of rubber modified asphalt – practical application (viscosity, penetration decrease; Road building and maintenance (11): 378-382 (2010).
7. Kisgyörgy L., Toth C., Geiger A.; Elastic modulus of asphalt with chemically stabilized rubber bitumen, Elastic modulus of asphalt with chemically stabilized rubber bitumen; Gradevinar (68): 533-541 (2016).
8. Ramez Al.-Mansob, Amiruddin Bin Ismail, Nur Izzi Md. Yusoff, Mojtaba Shojaei Baghini; Rheological Characteristics of Epoxidized Natural Rubber Modified Bitumen; Mechanics and Materials Vols. 505-506: 174-179 (2014).

9. Ibrahim M.R., Katman H.Y., Karim M.R., Kotling S., Mashaan N.S.; A review on the effect of crumb rubber addition to the rheology of crumb rubber modified bitumen; Materials Science and Engineering 3 (2013).
10. Radziszewski P., Sarnowski M., Król J., Kowalski K., Ruttmar I., Zborowski A.; Properties of rubber and mineral -rubber-modified asphalts, Warszawa (2017).
11. Radziszewski P., Kalabińska M., Piłat J.; Application of rubber modified asphalt in road building and maintenance; Drogownictwo (3): 73-77 (2001).
12. Koba H., Szydło A.; Influence of rubber modification on adhesion to rocks; Road building and maintenance (6): 198-201 (2010).
13. Radziszewski P., Plewa A.; Impact of rubber modified asphalt warming time on its viscoelastic properties; Drogownictwo (3): 86-49 (2003).
14. Pszczoła M., Jaczewski M., Szydłowski C., Dołżycki B., Judycki J.; Assessment of properties of low temperature mineral-asphalt mixtures with rubber and SBS modified asphalt properties; Conference on Modern Building Materials, Structures and Techniques; Wilno (2017).
15. Sun D., Li L.H.; Factors Affecting the Viscosity of Crumb Rubber-Modified Asphalt; Petroleum Science and Technology 28(15): 1555-1566 (2010).
16. Billiter T.C., Davison R.R., Glover C.J., Bullin J.A.; Physical properties of asphalt-rubber binder; Petroleum Science and Technology 15(3-4): 205-236 (1997).
17. Silvrano A. Dantas Neto, Marcio M. Farias, Paulo A.A. Pereira, Jorge C. Pais; Influence of crumb rubber gradation on asphalt-rubber properties; Asphalt Rubber 2006: proceedings; 679-692 (2006).
18. Venudharan V., Krishna Prapoorna Biligiri; Effect of crumb rubber gradation on asphalt binder modification: rheological evaluation, optimization and selection; Materials and Structures 50(2): 129 (2017).
19. Najzarek Z., Wełnowski J., Novel mechanochemical technology for valorization of waste tire rubber, Conference InterNanoPoland, Katowice (2016).
20. Kok B.V., Yilmaz M., Sengoz B., Sengur A., Avci E.; Investigation of complex modulus of base and SBS modified bitumen with artificial neurol networks; Expert Systems with Applications 37; 7775-7780 (2010).
21. Styk S., Błażejowski K.; Technology of asphalt layers ; Wydawnictwa Komunikacji i Łączności; Warszawa (2014).
22. Słowik M.; Selected issues of viscoelasticity of modified road asphalts containing elastometer SBS; Dissertation series; no. 508; Wydawnictwo Politechniki Poznańskiej; Poznań (2013).

AUTHOR INDEX